Exploring symmetric cones with quantum

AF497773

Kathleen M. Jimenez

Contents

List of Tables

List of Figures

Abstract

Quantum algorithms for optimization often achieve speedups in the problem dimension. Yet, their error dependence and sensitivity to scale makes it challenging to identify broad classes of optimization problems for which their is a clear advantage over classical algorithms. This dissertation is comprised of multiple projects spanning three parts that seek to reduce this gap.

Part I concerns quantum linear algebra. We provide a construction for implementing matrix arithmetic operations, such as Kronecker and Hadamard products, on a quantum computer. Then, we demonstrate how Iterative Refinement can be leveraged to exponentially improve the dependence on precision in the overall running time associated with classically solving linear systems of equations using quantum computers.

Part II concerns Quantum Interior Point methods (QIPMs) for Semidefinite and Second-order conic optimization. QIPMs attempt to speedup the bottleneck of the classical IPM by substituting the classical solution of the Newton linear system with the combined use of a quantum linear systems algorithm and quantum state tomography (with some classical computation between iterates). We present the first provably convergent QIPMs for Semidefinite and Second-order conic optimization, by properly symmetrizing the Newton linear system, and utilizing an orthogonal subspace representation of the search directions to ensure feasibility of the sequence of iterates generated by our algorithms. Using the techniques we develop in Part I, we obtain a speedup in the dimension over classical IPMs, and recover their polylogarithmic dependence on precision.

Part III concerns quantum frameworks for semidefinite optimization based on matrix exponentials and Gibbs states. By combining state of the art quantum computing algorithms with iterative refinement techniques we provide evidence of genuine end-to-end asymptotic speedups for solving the semidefinite approximation of Quadratic Unconstrained Binary Optimization (QUBO) problems. We also show that dequantizing our approach yields a classical algorithm that (up to polylogarithmic factors) runs in matrix multiplication time

Chapter 1

Introduction

Optimization constitutes an important class of problems in mathematics, and is concerned with determining the "best" solution in some quantitative sense. Though papers had been published on special instances of linear optimization far earlier [Fou26, Hit41, Kan39, dlVP11], it is widely accepted that the field of optimization began to crystallize as a component of military efforts during the Second World War. This also seems to lend an explanation as to why the field of mathematical optimization is often referred to as *operations research* to this day. A seminal moment came two years after the end of the war in 1947, when George Dantzig formally proposed a planning model based on linear inequalities, and a solution approach known as the *Simplex Method* [Dan48, DOW$^+$55]. Around roughly the same time, John von Neumann formalized a relationship between *Linear Optimization* (LO) problems and zero-sum games, and proposed the theory of linear optimization duality [vN47]. These developments subsequently permeated into other disciplines, facilitating the development of mathematical and economic theory

The Simplex method received constant improvements following its introduction, and to this day there remains a gap between its worst case theoretical running time, and the striking efficiency with which it can solve problems in practice. Researchers were determined to prove that the computational resources required to solve LO problems could be polynomially bounded in the size of the problem instance. It was not until 1979 that linear optimization problems were shown to be polynomial-time solvable, when Khachiyan introduced his ellipsoid method in [Kha80], taking inspiration from Shor's ellipsoid method for nonlinear optimization [Sho87].

After the Simplex method, Karmarkar's *projective method* was perhaps the second most important discovery of the 20th century when it was introduced in his seminal work [Kar84]. Though interior point-type algorithms had been studied much earlier by Dikin [Dik67, Dik74], Fiacco and McCormick [FM64] and Frisch [Fri54, Fri55, Fri56], Karmarkar proved that his method ran in polynomial time, and that it improved on the complexity of the ellipsoid method. Shortly thereafter, Nesterov and Nemirovskii introduced the paradigm of efficiently-computable self-concordant barrier functions [NN88, NN95], which extended the scope of IPMs to convex programs. This paved the way to study *Semidefinite Optimization* problems, which generalize

LO problems, and to this day IPMs remain the most prevalent algorithms for solving convex optimization problems in practice.

It should come as little surprise that the development of the field of optimization coincided with the advent of computer science and the birth of the digital computer. Similar to optimization, the underlying theory of contemporary computer science dates back to the World War II era, with Church and Turing formally defining algorithms in 1936 [Chu36, Tur36]. This notion of algorithms paved the way for the Church-Turing thesis, which established that a machine could utilize an algorithm to perform any possible computation, provided adequate computational resources. Turing then introduced the *universal Turing machine*, upon which most of theoretical computer science still relies. Famously, Turing would later use these ideas to break the Enigma codes used by the German army to pass along encrypted communication alongside other researchers at Bletchley Park.

In the late 20th century, Feynman and Main proposed a model of computation that would be constructed based on the theory of quantum mechanics [Fey82, Fey85, Man80]. In recent years, quantum computers have been shown to posses the potential to solve problems beyond the reach of classical computers. The first evidence of this phenomena dates back to the algorithm of Deutsch and Joszai [DJ92], which showed that quantum computers can achieve exponential speedups over deterministic classical devices. Following the work of Grover [Gro96], search algorithms were also shown to be quadratically faster in the quantum model of computation. Following further developments, researchers have begun to investigate whether quantum computers could be used to accelerate the solution of certain optimization problems.

At present, quantum devices are noisy, which complicates the direct quantization of classical algorithms for optimization. For example, polynomial complexity results for certain classes of interior point methods rely on the assumption that all of the computation being performed exactly using integer data. Thus, on one hand, it is important to consider how we can design quantum algorithms in which the computation at each iterate is, in some sense, robust to quantum noise.

Quantum algorithms for optimization often achieve speedups in the problem dimension [ANTZ21, BKL$^+$19, BKF22, vAG19, vAGGdW20]. Yet, their error dependence and sensitivity to scale makes it challenging to identify broad classes of optimization problems for which their is a clear advantage over classical algorithms. Iterative Refinement is a tool that was devised for computing extended precision solutions to linear systems of equations [GVL13] when classical computers were a few years ahead of the current state of quantum devices. These techniques were subsequently applied in the classical linear [GSW12, GSW16, GS20] and mixed-integer [ACDE07, CKSW11] optimization. By combining state of the art quantum computing algorithms with iterative refinement techniques this thesis seeks to provide evidence of genuine end-to-end asymptotic speedups for some optimization problems.

1.1 Overview

Following the introduction of preliminary material, this thesis is split into three parts. In **Part I**, we discuss how linear algebraic techniques, and applications such as linear systems solving can be performed on a quantum computer. **Part II** is devoted to studying how quantum algorithms for solving linear systems of equations can be used to accelerate Interior Point Methods for various classes of conic optimization. In **Part III** we study other quantum algorithms for solving Semidefinite Optimization problems that are based on *Gibbs sampling* techniques rather than quantum linear systems algorithms. In each part of this thesis, we leverage a classical computing technique known as *iterative refinement*, which allows one to construct highly accurate solutions to linear systems of equations and optimization problems using low-precision arithmetic.

In Chapter 2, we begin by defining some basic notation and concepts that are critical to this thesis. We then give a brief overview on convex optimization, with a particular focus on conic linear optimization. This is followed by an introduction on the basics of quantum computation, and discuss quantum data structures and subroutines such as quantum state tomography (among others), that are used throughout this work.

In Chapter 3, we provide an overview *quantum linear algebra* under the framework of block-encoded matrices [LC19, CGJ19, Gil19]. We also introduce some new techniques that build on the results of [CGJ19, Gil19], such as Kronecker and Hadamard products.

In Chapter 4 we show how one can solve classical linear systems of equations using quantum computers. While quantum linear systems algorithms (QLSAs) can prepare quantum states encoding the solution to a linear system

$$Ax = b$$

in time that is polylogarithmic in the problem dimension and the inverse of the precision to which we estimate the solution, the tomography procedure required to classically estimate the resulting state introduces polynomial dependence on the inverse precision. As a result, current quantum approaches for solving classical linear systems of equations are exponentially slower than inexact classical methods. To reconcile this issue, we devise an *Iterative Refinement* scheme that treats the combined the combined use of a QLSA and an optimal tomography algorithm from [vACGN22] as a quantum subroutine for classically solving linear systems. When provided access to quantum RAM (QRAM), the resulting scheme converges to an ϵ-precise solution (in ℓ_2-norm accuracy) of a given linear system in time

$$\mathcal{O}\left(\left(d\alpha\kappa_A^2 + ds \right) \cdot \mathrm{polylog}\left(d, \kappa_A, \frac{\|b\|}{\epsilon} \right) \right),$$

where d is the problem dimension, κ_A is an upper bound for $\|A^{-1}\|$, s is the maximum number of nonzeros found in any of the rows of A. Note that $\alpha \leq \|A\|_F$, where $\|\cdot\|_F$ is the Frobenius norm and $\|b\|$ is the Euclidean norm of the right hand side vector. The results of this chapter will be critical to the algorithms studied in **Part II**.

In Chapter 5 we present a *Quantum Interior Point Method* (QIPM) for solving Semidefinite Optimization (SDO) problems. By introducing a novel *Inexact-Feasible* IPM framework that redefines the primal and dual search directions as linear combinations of bases for the nullspace and row-space of the constraints matrices, our scheme ensures primal and dual feasibility are exactly satisfied by the sequence of iterates generated by the algorithm, and avoid issues that prevented the convergence of early QIPMs. We further demonstrate how one can use the techniques we introduce in Chapter 4 to exponentially speedup the solution of the so-called Newton linear system on a quantum computer. The resulting scheme can be used to solve SDO problems involving $n \times n$ matrices and $m = \mathcal{O}(n^2)$ constraints in time

$$\mathcal{O}\left(\sqrt{n}(n^3\kappa^2 + n^4) \cdot \mathrm{polylog}\left(n, \kappa, \frac{1}{\epsilon}\right)\right),$$

where κ is an upper bound on the condition number of the Newton linear systems encountered during the algorithm. We show that de-quantizing our scheme leads to a novel classical IF-IPM that runs in time

$$\mathcal{O}\left(n^{4.5}\kappa \cdot \mathrm{polylog}\left(n, \kappa, \frac{1}{\epsilon}\right)\right).$$

When applied to Linear Optimization problems, the worst-case running time of the IF-QIPM is

$$\mathcal{O}\left(\left(n^2\kappa^2 + n^{2.5}\right) \cdot \mathrm{polylog}\left(n, \kappa, \frac{1}{\epsilon}\right)\right),$$

and its classical counterpart is shown to have worst case running time

$$\mathcal{O}\left(n^{2.5}\kappa \cdot \mathrm{polylog}\left(n, \kappa, \frac{1}{\epsilon}\right)\right).$$

In Chapter 6, we specialize the techniques used in Chapter 5 to develop a QIPM for optimization over the Second Order Cone. Specifically, we devise a quantum algorithm that can solve SOCO problems to ϵ-optimality in time

$$\mathcal{O}\left(\sqrt{r}(n\kappa^2 + n^2) \cdot \mathrm{polylog}\left(n, \kappa, \frac{1}{\epsilon}\right)\right),$$

where r is the number of second order cones, and n is the number of variables.

In Chapter 7 we give classical and quantum algorithms for solving the Semidefinite Approximation of Quadratic Unconstrained Binary Optimization (QUBO) problems. While their dependence on other parameters suggests no overall speedup over classical methodologies, SDO solvers based on Gibbs sampling techniques may provide speedups in the low-precision regime. We exploit this fact to our advantage, and present an iterative refinement scheme for the Hamiltonian Updates algorithm of Brandão et al. [BKF22], to exponentially improve the dependence of their algorithm on precision. As a result, we obtain a classical algorithm to solve the semidefinite relaxation of Quadratically Unconstrained Binary Optimization problems (QUBOs) in matrix multiplication time. When provided access to quantum random access memory (QRAM), the quantum algorithm takes

$$\mathcal{O}\left(n^{1.5} \cdot \mathrm{polylog}\left(n, \|C\|_F, \frac{1}{\epsilon}\right)\right)$$

accesses to the QRAM and additional quantum gates (this is the standard way of describing complexity in the QRAM model of computation), plus $\mathcal{O}(ns)$ classical arithmetic operations — note that simply reading the cost matrix C takes $\mathcal{O}(ns)$ time. To the best of our knowledge, our classical and quantum algorithms are the fastest known algorithms in their respective model of computation for this class of problems, and our quantum algorithm provides a genuine asymptotic speedup over known classical solution methodologies, provided that we have access to QRAM. In the sparse-access input model (without QRAM), the algorithm takes $\widetilde{\mathcal{O}}_{n, \|C\|_F, \frac{1}{\epsilon}}\left(n^{1.5} s^{0.5+o(1)}\right)$ accesses to an oracle describing the coefficient matrix C and $\widetilde{\mathcal{O}}_{n, \|C\|_F, \frac{1}{\epsilon}}\left(n^{2.5} s^{0.5+o(1)}\right)$ additional gates, therefore yielding no quantum speedup (the quantum gate complexity is asymptotically larger than the classical complexity).

1.2   Papers

This dissertation is based on the following papers:

[ANTZ21]   B. Augustino, G. Nannicini, T. Terlaky and L.F. Zuluaga. "Quantum Interior Point Methods for Semidefinite Optimization". 2021. arXiv:2112.06025

[AMN^{+}22]   B. Augustino, M. Mohammadisiaroudi, G. Nannicini, T. Terlaky and L.F. Zuluaga. "Quantum Interior Point Methods for Second Order Cone Optimization". 2022.

[ANTZ23]   B. Augustino, G. Nannicini, T. Terlaky and L.F. Zuluaga. "Solving the semidefinite relaxation of QUBOs in matrix multiplication time, and faster with a quantum computer". 2023. arXiv:2301.04237

[MAF^{+}22]   M. Mohammadisiaroudi, B. Augustino, R. Fakhimi, G. Nannicini, T. Terlaky. "Exponentially More Precise Tomography for Quantum Linear System Solutions via Iterative Refinement". 2022

[AN23]   B. Augustino and G. Nannicini. "Quantum Linear Algebra". Floudas, Christodoulos A., and Panos M. Pardalos, eds. Encyclopedia of Optimization. Springer Science & Business Media, to appear.

Chapter 2

Preliminaries

2.1 Notation and terminology

We begin by fixing some basic notation and reviewing concepts that are fundamental to this thesis.

Basic notation

We distinguish the quantity a to the k-th power and the value of a at iterate k using round brackets, writing a^k and $a^{(k)}$ to denote these quantities, respectively.

A collection of elements is called a *set*. We typically define a set by writing

$$\mathcal{X} = \{x : x \text{ satisfies certain properties}\}.$$

If x is an element of a set $\mathcal{X}$, we write $x \in \mathcal{X}$. If $x \in \mathcal{X}_1$ implies that $x \in \mathcal{X}_2$ for two sets $\mathcal{X}_1$ and $\mathcal{X}_2$, then $\mathcal{X}_1$ is a *subset* of $\mathcal{X}_2$, and use the notation $\mathcal{X}_1 \subset \mathcal{X}_2$ to indicate this relationship. We say that $\mathcal{X}$ is a *subspace* if $\mathcal{X}$ contains every line that passes through two distinct points in it, i.e.,

$$\alpha_1 x + \alpha_2 \bar{x} \in \mathcal{X} \quad \text{for all } (x, \bar{x}) \in \mathcal{X} \times \mathcal{X} \text{ and } (\alpha_1, \alpha_2) \in \mathbb{R} \times \mathbb{R}.$$

A set is said to be *affine* if there exists $x \in \mathbb{R}^n$ and a subspace $\overline{\mathcal{X}} \subseteq \mathbb{R}^n$ with

$$\mathcal{X} = x + \overline{\mathcal{X}} = \{x + \bar{x} : \bar{x} \in \overline{\mathcal{X}}\}.$$

For sets $\mathcal{X}_1, \ldots, \mathcal{X}_n$, we denote their Cartesian product by $\prod_{i=1}^{n} \mathcal{X}_i = \mathcal{X}_1 \times \mathcal{X}_2 \times \cdots \times \mathcal{X}_n$.

We denote by $\log(x)$ the base-2 logarithm. The notation $\mathrm{poly}(x, y)$ refers to arbitrary polynomials in x and y, while $\mathrm{polylog}(x, y)$ is used for arbitrary polynomials in $\log(x)$ and $\log(y)$. For an element $x \in \mathbb{C}^d$, $\Re(x) \in \mathbb{R}^d$ is its real part. For a nonnegative integer n, write $[n]$ to represent the set of elements $\{1, \ldots, n\}$.

For any positive integer q, and binary strings $j, k \in \{0,1\}^q$, we denote by $j \oplus k$ the bitwise modulo 2 addition (element-wise XOR) of q-digit strings, defined as

$$j \oplus k = h$$

where $h \in \{0,1\}^q$ is the bitstring whose elements h_p are defined for $p = 1, \ldots, q$ as

$$h_p = \begin{cases} 0 & \text{if } j_p = k_p, \\ 1 & \text{otherwise.} \end{cases}$$

Linear algebra

We denote the i-th element of a vector $x \in \mathbb{R}^n$ by x_i for $i \in [n]$, and the ij-th element of a matrix $A \in \mathbb{R}^{m \times n}$ by A_{ij} for $i \in [m]$ and $j \in [n]$. To refer to the i-th row of a matrix A, we write $A_{i,\cdot}$ and write $A_{\cdot,j}$ when referring to its j-th column.

For a vector $v \in \mathbb{R}^n$, its *transpose* is the vector $v^\top \in \mathbb{R}^{1 \times n}$. The transpose of a matrix $A \in \mathbb{R}^{m \times n}$ is $A^\top \in \mathbb{R}^{n \times m}$, whose elements are defined by $(A^\top)_{ij} = A_{ji}$. When $A \in \mathbb{R}^{n \times n}$ and $A = A^\top$, we say that A is *symmetric* or alternatively write $A \in \mathcal{S}^n$, such that $\mathcal{S}^n$ is the space of $n \times n$ symmetric matrices in $\mathbb{R}^{n \times n}$. We denote the conjugate transpose of a complex matrix A by $A^\dagger$, and when $A = A^\dagger$, we say that A is *Hermitian*. One can observe that all real valued symmetric matrices are Hermitian.

We write $\langle \cdot, \cdot \rangle$ when referring to an arbitrary inner product on $\mathbb{R}^n$. The inner product $\langle \cdot, \cdot \rangle$ induces a norm on $\mathbb{R}^n$

$$\|x\| = \sqrt{\langle x, x \rangle}.$$

For instance, when $\langle \cdot, \cdot \rangle$ is the standard vector inner product (or, dot product) on $\mathbb{R}^n$, then the induced norm is the Euclidean norm. If $(x, \tilde{x}) \in \mathbb{R}^n \times \mathbb{R}^n$ satisfy $\langle x, \tilde{x} \rangle = 0$, then x and $\tilde{x}$ are *orthogonal*. A set of vectors $v_1, \ldots, v_n$ form an *orthonormal basis* for $\mathbb{R}^n$ if

$$\langle v_i, v_j \rangle = \delta_{ij} \quad \text{for all } i, j \in [n],$$

where δ_{ij} is the Kronecker delta. A linear operator $P : \mathbb{R}^n \mapsto \mathbb{R}^n$ is orthogonal if

$$\langle Px, P\tilde{x} \rangle = 0 \quad \text{for all } (x, \tilde{x}) \in \mathbb{R}^n \times \mathbb{R}^n.$$

If P is a matrix and its ambient space is equipped with an inner product, we can simply write $PP^\top = I$ (or equivalently, $P^\top P = I$).

For a matrix A, we refer to the set of all elements d that map A to 0 as the *nullspace* (or, *kernel*) of A, and we write $\text{Null}(A) = \{d : Ad = 0\}$. Conversely, we refer to the orthogonal complement of its nullspace as the *row-space* of A, i.e., $\mathcal{R}(A) = \text{Null}(A)^\perp$. Consequently, for any $x \in \text{Null}(A)$ and $y \in \mathcal{R}(A)$, we have $\langle x, y \rangle = 0$, where $\langle \cdot, \cdot \rangle$ is the inner product equipped by our algebraic space.

When $\mathbb{R}^n$ and $\mathbb{R}^m$ are equipped with inner products and $A : \mathbb{R}^n \mapsto \mathbb{R}^m$ is a linear operator, there exists a unique linear operator $A^* : \mathbb{R}^m \mapsto \mathbb{R}^n$ known as the *adjoint* of A, which satisfies

$$\langle Ax, y \rangle = \langle x, A^* y \rangle$$

for all $x \in \mathbb{R}^n$ and $y \in \mathbb{R}^m$. We say that a linear operator A is *self-adjoint* if $A = A^*$. Also note that if $\mathbb{R}^n$ and $\mathbb{R}^m$ are equipped with the Euclidean inner product and we can write A and A^* as matrices, then $A^* = A^\top$. One can observe that $\text{Null}(A)$ is the range space of A^*. Hence, Whenever A is surjective, its adjoint A^* is injective. In this case, $A^*(AA^*)^{-1}A$ is the orthogonal projection from $\mathbb{R}^n$ onto the range space of A^*, and the operator $I - A^*(AA^*)^{-1}A$ is the orthogonal projection onto $\text{Null}(A)$.

A problem that often arises in applications of linear algebra is that of determining scalar values λ for which there exists vectors $x \neq 0$ that satisfy

$$Ax = \lambda x.$$

There exist nontrivial solutions x to the above system if and only if there exists λ that solves the characteristic equation

$$\det(A - \lambda I) = 0,$$

where $\det(\cdot)$ denotes the determinant. The roots λ_i, $i = 1, \ldots, n$ of the characteristic equation are called the *eigenvalues* of A, and the vectors x which solve $Ax = \lambda x$ for a given eigenvalue λ are the *eigenvectors* of A. We denote the smallest and largest eigenvalues of A by $\lambda_{\min}(A)$ and $\lambda_{\max}(A)$. Whenever A is a symmetric matrix, its eigenvalues are real, and its eigenvectors are orthogonal. Letting $\lambda_1, \ldots, \lambda_n$ be the eigenvalues of $A \in \mathcal{S}^n$, and P be an orthogonal matrix whose columns are the eigenvectors of A, *spectral decomposition* of A is given by

$$A = P^\top \Lambda P,$$

where $\Lambda = \text{diag}(\lambda_1, \ldots, \lambda_n)$ and $P^\top = P^{-1}$.

A symmetric matrix is said to be *positive semidefinite* (*positive definite*) if all of its eigenvalues are nonnegative (positive). We let $\mathcal{S}^n_+$ ($\mathcal{S}^n_{++}$) represent the spaces of symmetric positive semidefinite (symmetric positive definite) matrices in $\mathbb{R}^{n \times n}$. For $A, B \in \mathcal{S}^n$, we write $A \succeq B$ ($A \succ B$) to indicate that the matrix $A - B$ is positive semidefinite (positive definite), i.e., $A - B \in \mathcal{S}^n_+$ ($A - B \in \mathcal{S}^n_{++}$). We list a number of equivalent characterizations of positive semidefiniteness below:

Theorem 2.1. *Let $A \in \mathcal{S}^n$. Then, the following are equivalent certificates of positive semidefiniteness:*

(a) $\lambda_{\min}(A) \geq 0$.

(b) The quadratic form $x^\top A x$ is nonnegative for all $x \in \mathbb{R}^n$.

(c) All principal minors of A are nonnegative.

(d) There exists a Cholesky factorization of A. For some matrix $L \in \mathbb{R}^{n \times k}$, we have

$$A = LL^{\top}.$$

We can also extend the above result to positive definite matrices.

Theorem 2.2. *Let $A \in \mathcal{S}^n$. Then, the following are equivalent certificates of positive definiteness:*

(a) $\lambda_{\min}(A) > 0$.

(b) The quadratic form $x^{\top} A x$ is positive for all $x \in \mathbb{R}^n$ with $x \neq 0$.

(c) All principal minors of A are positive.

(d) There exists a Cholesky factorization of A. For some nonsingular matrix $L \in \mathbb{R}^{n \times k}$, we have

$$A = LL^{\top}.$$

The *trace* of an $n \times n$ matrix is a linear mapping on $\mathbb{R}^{n \times n}$, and is defined as the sum of the diagonal elements:

$$\operatorname{tr}(A) := \sum_{i=1}^{n} A_{ii}.$$

The trace satisfies $\operatorname{tr}(A) = \operatorname{tr}(A^{\top})$ and $\operatorname{tr}(AB) = \operatorname{tr}(BA)$ where the latter is referred to as the cyclic property of the trace. For symmetric matrices, the trace is equal to the sum of the eigenvalues, i.e., $\operatorname{tr}(AB)$ is the sum of the eigenvalues of the matrix AB.

The singular values $\sigma_1, \ldots, \sigma_n$ of $A \in \mathbb{C}^{m \times n}$ are the square roots of the eigenvalues of $A^{\dagger} A \in \mathbb{C}^{n \times n}$. Though generally not the case, whenever A is positive semidefinite, its eigenvalues and singular values coincide. Suppose $\sigma_1 \leq \sigma_2 \leq \cdots \leq \sigma_n$ are the singular values of $A \in \mathbb{C}^{m \times n}$, then the *singular value decomposition* of A is defined as

$$A = W \Sigma V^{\dagger},$$

where $\Sigma = \operatorname{diag}(\sigma_1, \ldots, \sigma_n)$ and $W \in \mathbb{C}^{m \times m}$ and $W \in \mathbb{C}^{n \times n}$ are orthogonal matrices. In what follows, we denote the smallest and largest singular values of a matrix A by $\sigma_{\min}(A), \sigma_{\max}(A)$, respectively. The condition number of A is $\kappa_A = \frac{\sigma_{\max}(A)}{\sigma_{\min}(A)}$.

In what follows, $\|x\|_p$ denotes the ℓ_p-norm of a vector x. We occasionally omit the subscript when referring to the Euclidean norm of a vector, i.e., the special case of $p = 2$. The trace defines an inner product on $\mathbb{R}^{n \times n}$, defining

$$\langle A, B \rangle = \operatorname{tr}(A^{\top} B) = \sum_{i,j=1}^{n} A_{ij} B_{ij}$$

to be the *trace inner product*. The trace inner product generalizes the standard inner product defined for vector spaces of dimension n to the space of $n \times n$ matrices. Accordingly, the *Frobenius norm* of $A \in \mathbb{R}^{m \times n}$ is the extension of the Euclidean norm from vectors to matrices:

$$\|A\|_F := \sqrt{\sum_{i \in [m]} \sum_{j \in [n]} A_{ij}^2}.$$

For matrices, $\|A\|$ represents the *operator norm*, defined to be the largest singular value of A:

$$\|A\| := \max_{v \neq 0} \frac{\|Av\|}{\|v\|}.$$

The *trace norm* is defined as the sum of the singular values of A:

$$\|A\|_{\mathrm{tr}} := \mathrm{tr}\left(\sqrt{A^\dagger A}\right).$$

The *Schatten p-norm*, defined for a bounded linear operator A is given by

$$\|A\|_p := \left[\mathrm{tr}\left(|A|^p\right)\right]^{\frac{1}{p}},$$

where $|A| := (A^\dagger A)^{\frac{1}{2}}$. Notice that the trace and operator norms $\|\cdot\|_{\mathrm{tr}}$ and $\|\cdot\|$ are the Schatten-1 and Schatten-∞ norms, respectively, and the Frobenius norm $\|\cdot\|_F$ corresponds to the Schatten-2 norm.

Positive semidefinite matrices $A \in \mathcal{S}_+^n$ admit the useful identity

$$\|A\|_{\mathrm{tr}} = \mathrm{tr}(A) = \sum_{i \in [n]} \lambda_i(A),$$

where $\lambda_i(A)$ is the i-th eigenvalue of A. The equivalence is due to the fact that the singular values of A are equivalent to the eigenvalues of A whenever $A \in \mathcal{S}_+^n$.

When the dimension is clear from the context, we write e to refer to the vector of all ones, and use the notation e_i to refer to the i-th unit vector in the standard orthonormal basis $\{e_1, \ldots, e_n\}$ for $\mathbb{R}^n$. Moreover, $ee^\top$ is the all-ones matrix. For a vector $v \in \mathbb{R}^d$, $\mathrm{diag}(v) \in \mathbb{R}^{d \times d}$ is diagonal matrix with v along the diagonal. The matrix exponential $\exp(A)$, which is defined by the power series

$$\exp(A) = I + A + \frac{1}{2!}A^2 + \frac{1}{3!}A^3 + \cdots,$$

maps symmetric matrices to the space of symmetric positive definite matrices. Given the spectral decomposition $A = V\Lambda V^\top$, then $\exp(A) = V\exp(\Lambda)V^\top$, where

$$\exp(\Lambda) = \mathrm{diag}(\exp(\Lambda_{11}), \exp(\Lambda_{22}) \ldots, \exp(\Lambda_{nn})).$$

We denote by $A \otimes B$ the tensor product of two matrices, and $A^{\otimes n}$ is the n-fold tensor onto itself. For a matrix $A \in \mathbb{R}^{n \times n}$, $\mathrm{vec}(A)$ is the $n^2 \times 1$ vector given by

$$\mathrm{vec}(a) = (A_{11}, A_{21}, \ldots, A_{n1}, A_{12}, \ldots, A_{nn})^{\top}.$$

We define the symmetric analogue of the $\mathrm{vec}(\cdot)$ operator below.

Definition 2.1. *For $A \in \mathcal{S}^n$, $\mathrm{svec}(A) \in \mathbb{R}^{\frac{1}{2}n(n+1)}$ is given by*

$$\mathrm{svec}(A) = \left(A_{11}, \sqrt{2}A_{21}, \ldots, \sqrt{2}A_{n1}, A_{22}, \sqrt{2}A_{32} \ldots, \sqrt{2}A_{n2}, \ldots, A_{nn} \right)^{\top}.$$

Further, the operator $\mathrm{smat}(\cdot)$ is the inverse operator of $\mathrm{svec}(\cdot)$. That is,

$$\mathrm{smat}\left[\mathrm{svec}(A)\right] = A.$$

We also make use of the symmetric Kronecker product, defined below.

Definition 2.2. *Let $A, B \in \mathcal{S}^n$. Define the $\frac{n(n+1)}{2} \times n^2$ matrix V as:*

$$V_{(i,j),(k,l)} = \begin{cases} 1 & \text{if } i = j = k = l, \\ 1/\sqrt{2} & \text{if } i = j \neq k = l, \text{ or } i = l \neq j = k \\ 0 & \text{otherwise.} \end{cases}$$

Then, the symmetric Kronecker product $G \otimes_s K$ of G and K is defined as:

$$A \otimes_s B = \frac{1}{2}V \left(A \otimes B + B \otimes A \right) V^{\top}.$$

Computational cost

When discussing the computational complexity of quantum algorithms we normally express the cost in terms of the number of calls to some input oracle. Unless otherwise specified, the gate complexity is at most a poly-logarithmic factor larger than the stated oracle complexity. The meaning of "input oracle access" depends on the input model:

- For the sparse-oracle access model, it refers to a query to the oracle describing the input data.

- For the QRAM model, it refers to the number of accesses to QRAM. The algorithms we study in this thesis only require classical write access to the QRAM, i.e., we do not write in superposition.

It is straightforward to translate each of these oracle costs into a running time in the standard gate model without QRAM, by considering the cost of implementing each oracle.

We define $\mathcal{O}(\cdot)$ as

$$f(x) = \mathcal{O}(g(x)) \iff \exists \ell \in \mathbb{R}, c \in \mathbb{R}_{+}, \text{ such that } f(x) \leq cg(x) \quad \forall x > \ell.$$

We write $f(x) = \Omega(g(x)) \iff g(x) = \mathcal{O}(f(x))$, and

$$f(x) = \Theta(g(x)) \iff f(x) = \mathcal{O}(g(x)) \text{ and } f(x) = \Omega(g(x)).$$

We also define $\tilde{\mathcal{O}}(f(x)) = \mathcal{O}(f(x) \cdot \text{polylog}(f(x)))$ and when the function depends poly-logarithmically on other variables we write

$$\tilde{\mathcal{O}}_{a,b}\left(f(x)\right) = \mathcal{O}(f(x) \cdot \text{polylog}(a, b, f(x))).$$

The functions $\tilde{\Omega}_{a,b}(\cdot)$ and $\tilde{\Theta}_{a,b}(\cdot)$ are defined similar to $\tilde{\mathcal{O}}_{a,b}(\cdot)$.

2.2 Convex Optimization

The focus of *convex optimization* is to derive conditions under which optimal solutions exist for constrained problems of the form:

$$\text{minimize } f(x)$$
$$\text{subject to } x \in \mathcal{X}, \quad g_j(x) \leq 0, \quad j = 1, \ldots, m.$$

Here, $f(\cdot)$ is the *objective function*, and the *objective value* associated with a solution x is $z = f(x)$. If $x \in \mathcal{X}$ satisfies $g_j(x) \leq 0$, for all $j = 1, \ldots, m$, we say that x is *feasible*. The set $\mathcal{X}$ is understood to be *convex*, meaning that for any two points $x \in \mathcal{X}$ and $\bar{x} \in \mathcal{X}$, any point determined by a *convex combination* of them is also a member of the set $\mathcal{X}$, i.e.:

$$\alpha x + (1 - \alpha)\bar{x} \in \mathcal{X},$$

for all $\alpha \in [0, 1]$. The notion of convex combinations readily generalizes to higher dimensions; x is a convex combination of $(x_1, \ldots, x_n)^\top \in \mathbb{R}^n$ if

$$x = \sum_{i=1}^{n} \alpha_i x_i,$$

with $\sum_{i=1}^{n} \alpha_i = 1$. The *convex hull* of a set $\mathcal{X}$ is the intersection of all convex sets containing $\mathcal{X}$. If $\mathcal{X}$ is convex, then its convex hull is just the set itself, i.e., $\text{conv}(\mathcal{X}) = \mathcal{X}$.

The definition of convexity for sets also naturally extends to functions. A function $f : \mathcal{C} \subset \mathbb{R}^n \mapsto \mathbb{R}$ is convex if

$$f(\alpha x + (1 - \alpha)\bar{x}) \leq \alpha f(x) + (1 - \alpha)f(\bar{x})$$

for all $(x, \bar{x}) \in \mathcal{C} \times \mathcal{C}$ and $\alpha \in [0, 1]$. In other words, the function value $f(\alpha x + (1 - \alpha)\bar{x})$ *underestimates* the linear interpolation $\alpha f(x) + (1 - \alpha)f(\bar{x})$. The set of points lying above the graph of a function $f(\cdot)$, known as the epigraph of f, provides another avenue for determining convexity. Namely, a function $f(\cdot)$ is convex if and only if

$$\text{epi}(f) = \{(x, w) : x \in \mathcal{C}, w \in \mathbb{R}, f(x) \leq w\}.$$

is a convex set. If $-f(\cdot)$ is convex, then $f(\cdot)$ is *concave*. Examples of convex functions include linear functions, ℓ_p-norms, the $\max(\cdot)$ operator, as well as the $\log(\det(X))$ for matrices $X \in \mathbb{R}^{n \times n}$ (which plays a crucial role in Interior Point Methods for semidefinite optimization).

Some of the most crucial concepts in convexity and optimization involve *hyperplanes*, which are sets of the form

$$\left\{ x \in \mathbb{R}^n : a^\top x = b \right\},$$

where $b \in \mathbb{R}$ and $a \in \mathbb{R}^n$ is called a *normal vector* of the hyperplane, with $a \neq 0$. Note that a is orthogonal to the hyperplane itself, passing through the origin. A (closed) *halfspace* is a set of the form

$$\left\{ x \in \mathbb{R}^n : a^\top x \leq b \right\},$$

and when the inequality is strict, we say that the halfspace is *open*. Halfspaces are convex, but not affine. Using halfspaces, we can define *polyhedral sets*, which when defined in $\mathbb{R}^n$, take the form

$$\left\{ x : a_i^\top x \leq b_i \text{ for all } i \in [m] \right\},$$

where $a_i \in \mathbb{R}^n$ and $b_i \in \mathbb{R}$ for all $i \in [m]$[1]. If the polyhedron is bounded, it may be referred to as a *polytope*.

2.3 Conic optimization and duality

Of crucial interest to this thesis are sets known as *cones*. A set $\mathcal{K} \subseteq \mathbb{R}^n$ is a cone if

$$\alpha x \in \mathcal{K}$$

for all $x \in \mathcal{K}$ and $\alpha \geq 0$. Letting $\alpha_1, \alpha_2 \in \mathbb{R}_+$, we say that $\mathcal{K}$ is a *convex cone* if

$$\alpha_1 x + \alpha_2 \bar{x} \in \mathcal{K}$$

for all $(x, \bar{x}) \in \mathcal{K} \times \mathcal{K}$. It can be easily verified that convex cones are convex sets.

While not all cones are convex, many important ones are. These include, but are not limited to the nonnegative orthant $\mathbb{R}^n_+$, the cone of $n \times n$ symmetric positive semidefinite matrices $\mathcal{S}^n_+$, as well as the second order (or, *Lorentz*) cone:

$$\left\{ x \in \mathbb{R}^n : x_1^2 - \sum_{i=2}^{n} x_i^2 \geq 0, x_1 \geq 0 \right\}.$$

---

[1] It is important to assume $m < \infty$.

In fact, optimization over these three cones gives rise to the classes of Linear, Semidefinite, and Second-Order Conic Optimization, respectively.

We say that $\mathcal{K}$ is a *pointed cone* if $x, -x \in \mathcal{K}$ implies that $x = 0$. If $\mathcal{K} \subset \mathbb{R}^n$ is a pointed convex cone, then it defines a partial ordering on $\mathbb{R}^n$; for $(x, \bar{x}) \in \mathbb{R}^n \times \mathbb{R}^n$:

$$x \succeq_\mathcal{K} \bar{x} \iff x - \bar{x} \in \mathcal{K}.$$

Letting $\text{int}(\mathcal{K})$ denote the interior of $\mathcal{K}$, then $x \succ_\mathcal{K} \bar{x}$ means $x - \bar{x} \in \text{int}(\mathcal{K})$.

A cone $\mathcal{K} \subset \mathbb{R}^n$ is *proper* if it is closed, pointed and has a nonempty interior. One can express the third property by writing $\text{int}(\mathcal{K}) \neq \emptyset$ or equivalently $\mathcal{K} + (-\mathcal{K}) = \mathbb{R}^n$. Throughout this thesis, we assume $\mathcal{K}$ is a closed, pointed convex cone in $\mathbb{R}^n$ with a nonempty interior.

The *dual cone* of a cone $\mathcal{K}$ is defined as

$$\mathcal{K}^* = \left\{ s : \langle s, x \rangle \geq 0 \text{ for every } x \in \mathcal{K} \right\}.$$

Note that the definition of the dual cone is dependent on the choice of inner product. A cone $\mathcal{K}$ is *self-dual* if $\mathcal{K} = \mathcal{K}^*$.

The nonnegative orthant $\mathbb{R}^n_+$, the cone of $n \times n$ symmetric positive semidefinite matrices $\mathcal{S}^n_+$, and the second order cone are all self-dual. In fact, these three cones are all *symmetric* (or, self-scaled). A cone $\mathcal{K}$ is symmetric if it is self-dual, and homogeneous, i.e., for any $(x, \bar{x}) \in \text{int}(\mathcal{K})$, there exists a linear transformation M such that $M(x) = \bar{x}$ and $M(\mathcal{K}) = \mathcal{K}$.

Let $c \in \mathbb{R}^n$, $b \in \mathbb{R}^m$ be given vectors and $A : \mathbb{R}^n \mapsto \mathbb{R}^m$ be a linear operator. Then, the *primal* Conic Linear Optimization (CLO) problem can be defined in standard form as follows:

$$z_P = \inf_x \left\{ \langle c, x \rangle : Ax = b, x \succeq_\mathcal{K} 0 \right\}. \tag{CLO-P}$$

In this context, we say that c is the objective vector and b is the right hand side vector. If $x \succeq_\mathcal{K} 0$ satisfies $Ax = b$, we say that x is *primal feasible*, and $\langle c, x \rangle$ is the *objective value* associated with this solution. If $x \succ_\mathcal{K} 0$ with $Ax = b$, then x is said to be *strictly feasible*. Note that we have written "inf" rather than "min" even though we assume $\mathcal{K}$ is closed.

For every primal problem, there is a corresponding *dual* problem, whose construction is motivated by the following question:

How can we obtain the best lower bound for the optimal objective value of the primal problem?

Suppose $x \in \mathcal{K}$ satisfies $Ax = b$. Then, for any $y \in \mathbb{R}^m$, we also have $\langle y, b - Ax \rangle = \langle y, 0 \rangle = 0$. Combining this with the definition of the adjoint A^* yields

$$\langle c, x \rangle = \langle c, x \rangle + 0 = \langle c, x \rangle + \langle y, b - Ax \rangle = \langle y, b \rangle + \langle c - A^* y, x \rangle.$$

Suppose that we require $c - A^* y \in \mathcal{K}^*$. Then, recalling that $x \in \mathcal{K}$, we may write

$$\langle y, b \rangle + \langle c - A^* y, x \rangle \geq \langle y, b \rangle,$$

as $\langle c - A^\top y, x \rangle \geq 0$ by the definition of $\mathcal{K}^*$. Therefore, the best lower bound we can hope for would be obtained by solving the optimization problem

$$z_D = \sup_{y} \left\{ \langle b, y \rangle : c - A^* y \succeq_{\mathcal{K}^*} 0 \right\},$$

and by introducing a *dual slack variable* $s \in \mathbb{R}^n$, the above problem can also be written in standard form:

$$z_D = \sup_{(y,s)} \left\{ \langle b, y \rangle : A^* y + s = c, s \succeq_{\mathcal{K}^*} 0 \right\}. \tag{CLO-D}$$

This is the *dual* CLO problem.

By design, for any primal-dual pair of CLO problems we always have

$$\langle b, y \rangle \leq \langle c, x \rangle,$$

for any primal feasible x and dual feasible y, which is a property known as *weak duality*. If the relationship holds at equality, i.e.,

$$\langle b, y \rangle = \langle c, x \rangle,$$

where x is a primal optimal solution and y is a dual optimal solution, we say that *strong duality* holds. While weak duality holds without exception, the same cannot be said for strong duality, though we remark that strong duality will always hold for linear optimization problems when the optimal objective value is finite.

Next, we discuss how specific choices for $\mathcal{K}$ give rise to some of the most common classes of convex optimization.

2.3.1 Linear optimization

In Linear optimization (LO), one seeks to determine the optimal value of a linear function that is subject to linear constraints on the variables which are defined as elements nonnegative orthant, i.e.,

$$\mathbb{R}^n_+ = \left\{ x \in \mathbb{R}^n : x_i \geq 0 \; \forall i \in [n] \right\}.$$

The nonnegative orthant lies in the Euclidean space $\mathbb{R}^n$ and is equipped with the standard inner product, and accordingly, LO is a special case of CLO that arises upon setting $\mathcal{K} = \mathbb{R}^n_+$. Though LO problems may have either equality or inequality constraints, they can always be re-written

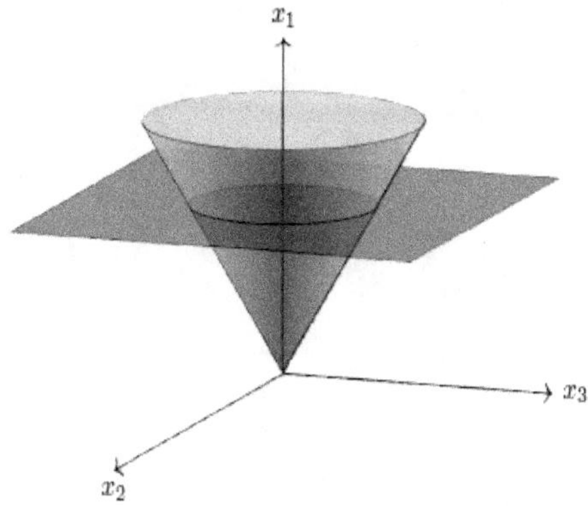

Figure 2.1: A second-order cone in $\mathbb{R}^3$ intersected with an affine space.

in standard form, in which case the *primal* LO problem is defined as

$$
\begin{aligned}
\text{minimize} \quad & c^\top x \\
\text{subject to} \quad & Ax = b \\
& x \in \mathbb{R}^n_+,
\end{aligned}
\tag{LO-P}
$$

and its associated dual problem is of the form

$$
\begin{aligned}
\text{maximize} \quad & b^\top y \\
\text{subject to} \quad & A^\top y + s = c \\
& y \in \mathbb{R}^m, \ s \in \mathbb{R}^n_+.
\end{aligned}
\tag{LO-D}
$$

LO is perhaps the most commonly studied class of optimization, due to the fact that it possess a number of attractive qualities. Linear models are simple to construct while still providing an accurate enough of the problem at hand, and applications of LO can be found in areas such as engineering, statistics, as well as economic theory. Moreover, duality results are generally stronger for LO when compared to more complicated classes of CLO.

2.3.2 Second-order conic optimization

Second-order conic optimization (SOCO) corresponds to the case in which one seeks to minimize a linear objective function subject to a feasible region defined by the intersection of an affine space with the Cartesian product of a finite number of second-order cones. Thus SOCO problems generalize LO problems.

Interchangeably referred to as the *Lorentz* or "ice-cream" cone (see, e.g., Figure 2.1), a second-order cone can be defined as

$$
\mathcal{K} = \left\{ x \in \mathbb{R}^n : x_1^2 - \sum_{i=2}^{n} x_i^2 \geq 0, x_1 \geq 0 \right\}.
$$

The *primal* SOCO problem is defined as

$$
\begin{aligned}
\text{minimize} \quad & c^\top x \\
\text{subject to} \quad & Ax = b \\
& x \in \mathcal{K},
\end{aligned}
\tag{SOCO-P}
$$

and its associated dual problem is of the form

$$
\begin{aligned}
\text{maximize} \quad & b^\top y \\
\text{subject to} \quad & A^\top y + s = c \\
& y \in \mathbb{R}^m, \ s \in \mathcal{K}.
\end{aligned}
\tag{SOCO-D}
$$

Unlike LO, strong duality may fail for SOCO. To illustrate this phenomena, consider the following SOCO problem:

$$
\begin{aligned}
\min_{x} \quad & x_2 \\
\text{subject to} \quad & x_1 - x_3 = 0, \\
& \sqrt{x_2^2 + x_3^2} \le x_1,
\end{aligned}
$$

whose dual problem is given by

$$
\begin{aligned}
\max_{(y,s)} \quad & 0 \cdot y \\
\text{subject to} \quad & y + s_1 = 0, \\
& s_2 = 1, \\
& -y + s_3 = 0, \\
& \sqrt{s_2^2 + s_3^2} \le s_1.
\end{aligned}
$$

Clearly, the set of primal optimal solutions is $\{x \in \mathbb{R}^3 : x_1 = x_3 \ge 0, x_2 = 0\}$, while the dual problem is *infeasible*.

2.3.3 Semidefinite optimization

Choosing $\mathcal{K}$ to be the cone of positive semidefinite matrices $\mathcal{S}_+^n$ gives rise to Semidefinite Optimization (SDO).

Letting $A_1, \ldots, A_m, C \in \mathcal{S}^n$ and $b \in \mathbb{R}^m$ be given data, the primal SDO problem in its

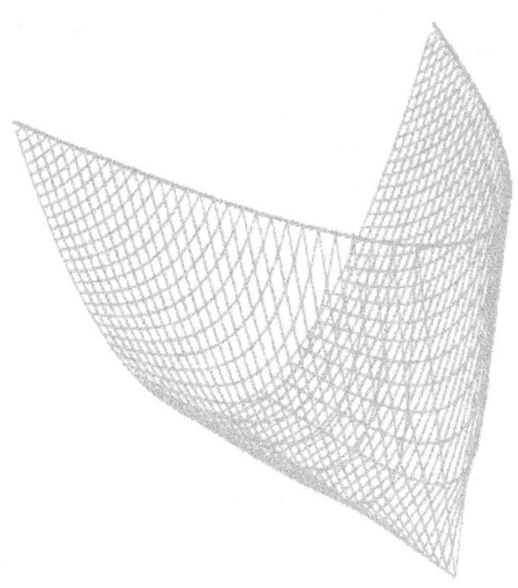

Figure 2.2: A positive semidefinite cone.

standard form is given by:

$$\begin{aligned}
\inf \quad & \operatorname{tr}(CX) \\
\text{subject to} \quad & \operatorname{tr}(A_i X) = b_i \quad \forall i \in [m] \\
& X \in \mathcal{S}_+^n.
\end{aligned} \qquad \text{(SDO-P)}$$

That is, we seek to determine a matrix variable X that minimizes a linear objective function, where X is required to be positive semidefinite and lie in the affine subspace

$$\mathcal{F} = \{X \in \mathcal{S}^n : \operatorname{tr}(A_i X) = b_i \quad \forall i \in [m]\}.$$

If $X \in \mathcal{S}_+^n \cap \mathcal{F}$, we say that X is *feasible*, and if in fact $X \in \mathcal{S}_{++}^n \cap \mathcal{F}$, then X is *strictly feasible*. The dual SDO problem associated with (SDO-P) is of the form

$$\begin{aligned}
\sup \quad & b^\top y \\
\text{subject to} \quad & \sum_{i=1}^m A_i^\top y_i - S = C \\
& y \in \mathbb{R}^m, \ S \in \mathcal{S}_+^n,
\end{aligned} \qquad \text{(SDO-D)}$$

and $S = C - \sum_{i=1}^m A_i^\top y_i$ is the *dual slack matrix*.

SDO problems generalize LO problems, which are SDO problems in which each of the input matrices $A_1, \ldots, A_m, C \in \mathcal{S}^n$ are all diagonal. Likewise, second-order cones admit a matrix representation which illustrates how SOCO can be cast a special case of SDO. For any $x \in \mathbb{R}^n$, consider the arrowhead matrix

$$\mathrm{Arw}(x) = \begin{pmatrix} x_1 & x_{2:n} \\ x_{2:n}^\top & x_1 I_{n-1} \end{pmatrix}, \tag{2.1}$$

where $x_{2:n} \equiv (x_2, x_3, \ldots, x_n)$ and I_{n-1} denotes the identity matrix of order $n-1$. One can easily verify that $x \in \mathcal{K}$ if and only if $\mathrm{Arw}(x)$ is positive semidefinite.

It should therefore come as no surprise that SDO can elicit stronger results than LO and SOCO, and provides a modelling framework that reaches beyond their scope. On the other hand, as in the case of SOCO, duality results are weaker for SDO than LO; while weak duality holds unconditionally for a primal feasible X and dual feasible (y, S), strong duality does not always hold. However, strong duality will hold if both the primal and dual are feasible, with at least one of them being strictly feasible.

2.4 Interior Point Methods

The prevailing approach for solving conic optimization problems are *Interior Point Methods* (IPMs), which are a central topic of this thesis.

IPMs find their origin in *interior penalty methods*. Consider a generic constrained convex optimization problem

$$\min_{x \in \mathcal{X}} \langle c, x \rangle \tag{2.2}$$

where $\mathcal{X} \subset \mathbb{R}^n$ is a closed convex set. At one point in time, it was accepted that solving smooth convex optimization problems was easy, and so to solve a constrained problem, one should simply recast it as an unconstrained problem [BTN01]. Namely, one could introduce a *penalty function* $F(x)$ for $\mathcal{X}$, and solve (2.2) by applying Newton's method to a sequence of problems of the form

$$\min_{x \in \mathrm{int}(\mathcal{X})} \langle c, x \rangle + \mu F(x), \tag{2.3}$$

where μ is the nonnegative penalty constant. By design, $F(x)$ would be chosen to diverge to ∞ upon approaching the boundary of $\mathcal{X}$, in which case the only "constraint" present in (2.3) is that $F(x)$ is only defined on $\mathrm{int}(\mathcal{X})$. The hope was that the quadratic (local) convergence rate of Newton's method would lead allow one to quickly solve these penalty problems, and hence, the original problem of interest.

However, as the concepts of convergence and computational complexity of optimization algorithms began to take shape, a gap between theory and practice became evident. The interior penalty framework, though intuitive and flexible, was theoretically inadequate. The convergence guarantees of Newton's method are dependent on a number of problem-specific characteristics (e.g., smoothness, strong convexity), and that the algorithm is initialized to a starting point which is in some sense, close to the optimal solution. Unfortunately, researchers were left with more questions than answers (how to choose a starting point? how to define μ?). Nothing of consequence could be said about solving general classes of optimization problems in polynomial time, and a new approach was needed.

It turns out that the universality of the penalty method was part of the problem. Researchers were able to show that a polynomial time algorithm could be obtained by introducing more structure. Namely, this was found to be precisely the case when $F(\cdot)$ was a *self-concordant barrier function*. The paradigm of self-concordant barrier functions was introduced by Nesterov and Nemirovskii [NN88, NN95] and the so-called IPM revolution was kicked off. At a high level, one can think of self-concordant function as one whose third derivative can be upper bounded using its second derivative. Under some mild conditions (which will hold under an interior point framework), self-concordance ensures that taking the second order Taylor series expansion of our objective in (2.3) yields a rather accurate approximation.

One can view IPMs as a homotopy approach for Newton's method. IPMs are initialized to a strictly feasible point $x^{(0)} \in \text{int}\,(\mathcal{X})$, and approximately track an analytic curve known as the *central path* towards optimality. Specifically, letting $\mu > 0$ and defining $F : \text{int}\,(\mathcal{X}) \mapsto \mathbb{R}$, a generic path-following scheme solves a sequence of *barrier problems* of the form

$$x^*(\mu) := \operatorname*{argmin}_{x \in \text{int}\,(\mathcal{X})} \quad \langle c, x \rangle + \mu F(x) \tag{2.4}$$

using Newton's method. The value of μ is decreased in each iteration, and so closely tracing the central path $\{x^*(\mu) : \mu > 0\}$, we approach an optimal solution to (2.2) as $\mu \to 0$. The use of the barrier function F in (2.4) ensures that we are always in the region of rapid local convergence enjoyed by Newton's method. A visualization of an IPM applied to a linear optimization problem is provided in Figure 2.3.

It is also standard in the literature to define the central path as the set of minimizers associated with

$$x^*(t) := \operatorname*{argmin}_{x \in \text{int}(\mathcal{X})} \quad t\langle c, x \rangle + F(x), \tag{2.5}$$

in which case the optimal solution is reached upon tracing the central path $\{x^*(t) : t > 0\}$ as $t \to \infty$. Note however, there is no real difference between (2.4) and (2.5). Indeed, taking $\mu = 1/t$ and applying Newton's method to (2.4) yields the same sequence of minimizers to that one obtains upon applying Newton's method to (2.5).

2.5　A gentle introduction to quantum computing

In this section, we provide a basic overview of fundamental concepts in quantum computing that are used throughout this thesis. For a rigorous introduction, we refer the reader to the textbook of Nielsen and Chuang [NC02], as well as the lecture notes of de Wolf [dW19] and Lin [Lin22]. For those without a background in physics or quantum mechanics, we refer the reader to the introduction from Nannicini [Nan20], which provides a "physics-free" exposition that abstracts away from these concepts.

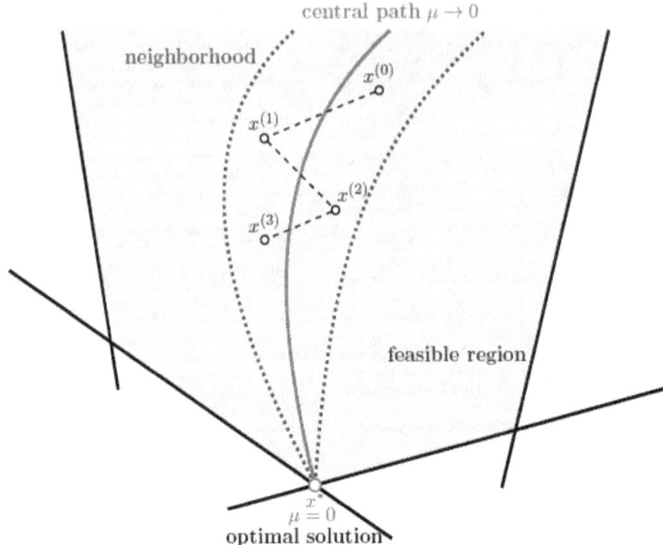

Figure 2.3: Iterates of an interior point method, visualized in its application to linear optimization.

Generally speaking, the quantum model of computation shares some notable similarities with its classical counterpart. However, rather than performing operations on vectors and matrices, we consider a *quantum state*, which following some predefined initialization, is maintained in a *quantum register*, and evolves under *unitary transformations*. When all of the computation has been completed, we can obtained some information on the state of the quantum register via *measurements*. We discuss these concepts, as well as useful quantum subroutines in detail next.

2.5.1   Qubits and quantum states

In what follows, we describe the *quantum circuit model*. We consider quantum systems that can be in a finite number N different (and mutually exclusive) classical states, which we denote by $|0\rangle, \ldots, |N-1\rangle$. Here, the states $|0\rangle, \ldots, |N-1\rangle$ form an orthonormal basis of an N-dimensional *Hilbert Space* (i.e., a complex Euclidean vector space equipped with an inner product). We refer to $|0\rangle, \ldots, |N-1\rangle$ as the *computational basis states*, and state $|i\rangle$ can be viewed as the quantum analogue to e_i, the i-th unit vector in the standard orthonormal basis $\{e_1, \ldots, e_N\}$ for $\mathbb{R}^N$.

The underlying principle of quantum mechanics is that, in contrast with classical systems, quantum states can be in what is known as a *superposition* of states. If we assign an *amplitude* $\alpha_i \in \mathbb{C}$ to each state $|i\rangle$, a *pure state* is any state of our quantum system that can be written as

$$|\phi\rangle = \sum_{i=0}^{N-1} \alpha_i |i\rangle = \begin{pmatrix} \alpha_0 \\ \vdots \\ \alpha_{N-1} \end{pmatrix},$$

with $\sum_{i=0}^{N-1}|\alpha_i|^2 = 1$. The component of the amplitude that corresponds to the argument of the complex number is called a *phase*, and is of the form $\exp(i\theta)$, where i is the imaginary unit.

Following *Dirac notation*, we denote the conjugate transpose of $|\phi\rangle$ by $\langle\phi|$. Observe that $|\phi\rangle$ (referred to as a *ket*) corresponds to a column vector, while $\langle\phi|$ (referred to as a *bra*) is a row vector, such that for two states $|\phi\rangle$ and $|\psi\rangle$, the expression $\langle\psi|\phi\rangle$ defines an inner product. Conversely, $|\phi\rangle\langle\psi|$ yields their outer product, and hence $|\phi\rangle\langle\phi|$ corresponds to a rank-1 matrix. Note that for quantum states we always have $\||\phi\rangle\| = \sqrt{\langle\phi|\phi\rangle} = 1$.

Hilbert spaces can also be combined through tensor products. Letting $|0\rangle, \ldots, |N-1\rangle$ be an orthonormal basis for a Hilbert space $\mathcal{H}_A$, and $|0\rangle, \ldots, |M-1\rangle$ be an orthonormal basis for a Hilbert space $\mathcal{H}_B$, then the composite system $\mathcal{H}_A \otimes \mathcal{H}_B$ gives rise to an NM-dimensional Hilbert space. In this case, $\mathcal{H}_A \otimes \mathcal{H}_B$ is spanned by the set of states

$$\left\{ |i\rangle \otimes |j\rangle : i \in \{0, \ldots, N-1\}, \ j \in \{0, \ldots, M-1\} \right\},$$

and states from the composite system can be written as

$$\sum_{i=0}^{N-1} \sum_{j=0}^{M-1} \alpha_{ij}|i\rangle \otimes |j\rangle.$$

Qubits are the quantum analogue of classical *bits*. Bits are the most rudimentary non-trivial classical system, and any bit has two possible states:

$$|0\rangle = \begin{pmatrix} 1 \\ 0 \end{pmatrix} \text{ and } |1\rangle = \begin{pmatrix} 0 \\ 1 \end{pmatrix}.$$

While classical computers have registers comprised of bits, a quantum computer possess a single quantum register that is made up of qubits. Now, a single qubit can be in any superposition

$$\alpha_0|0\rangle + \alpha_1|1\rangle$$

with $|\alpha_0|^2 + |\alpha_1|^2 = 1$. Upon generalizing this notion to q-qubit systems, our register can be in any superposition of 2^q classical states:

$$\alpha_0|0\rangle + \alpha_1|1\rangle + \cdots + \alpha_{2^q-1}|2^q-1\rangle, \quad \sum_{i=0}^{2^q-1}|\alpha_i|^2 = 1.$$

We may therefore define the state of a q-qubit quantum register as a unit vector in the composite system that arises upon taking a q-fold tensor product on $\mathbb{C}^2$:

$$(\mathbb{C}^2)^{\otimes q} = \underbrace{\mathbb{C}^2 \otimes \cdots \otimes \mathbb{C}^2}_{q \text{ times}}.$$

In other words, a q-qubit quantum register can be in any of 2^q states.

Along this line, we can define a basis for $(\mathbb{C}^2)^{\otimes q}$ using the following 2^q vectors

$$\underbrace{|00\cdots00\rangle}_{q\text{ digits}} = \underbrace{|0\rangle \otimes \cdots \otimes |0\rangle \otimes |0\rangle}_{q\text{ times}}$$

$$\underbrace{|00\cdots00\rangle}_{q\text{ digits}} = \underbrace{|0\rangle \otimes \cdots \otimes |0\rangle \otimes |1\rangle}_{q\text{ times}}$$

$$\vdots$$

$$\underbrace{|11\cdots11\rangle}_{q\text{ digits}} = \underbrace{|1\rangle \otimes \cdots \otimes |1\rangle \otimes |1\rangle}_{q\text{ times}}.$$

We will refer to the state of a q-qubit register as a q-qubit state. A quantum state $|\phi\rangle \in (\mathbb{C}^2)^{\otimes q}$ is said to be a *product state* (or alternatively, is *separable*) if it can be written as the tensor product of q 1-qubit states:

$$|\phi\rangle = |\phi_1\rangle \otimes \cdots \otimes |\phi_q\rangle,$$

otherwise, we say that $|\phi\rangle$ is *entangled*. It is typically assumed that the initial state of a quantum circuit is the q-qubit zero state $|0\rangle$.

2.5.2   Unitary evolution and gates

In applying operations to quantum states, we seek to change $|\phi\rangle$ to a state

$$|\psi\rangle = \beta_0|0\rangle + \beta_1|1\rangle + \cdots + \beta_{N-1}|N-1\rangle.$$

However, the transformations that we wish to apply must satisfy certain properties in order to ensure that we do not lose the basic properties of a quantum state upon doing so. The quantum model of computation allows for *linear* transformations. If we consider $|\phi\rangle = (\alpha_0, \ldots, \alpha_{N-1})^\top$, then, evolving $|\phi\rangle$ to $|\psi\rangle$ corresponds to applying an $N \times N$ complex matrix U to $|\phi\rangle$:

$$U \begin{pmatrix} \alpha_0 \\ \vdots \\ \alpha_{N-1} \end{pmatrix} = \begin{pmatrix} \beta_0 \\ \vdots \\ \beta_{N-1} \end{pmatrix}.$$

Accordingly, we may write $|\psi\rangle = U|\phi\rangle = U\left(\sum_{i=0}^{N-1} \alpha_i|i\rangle\right) = \sum_{i=0}^{N-1} \alpha_i U|i\rangle$.

Since $|\psi\rangle$ is a quantum state, we must have $\sum_{i=0}^{N-1}|\beta_i|^2 = 1$. This further restricts the set of mappings at our disposal; not only should our mappings be linear, but they must also preserve the norm of vectors. In other words, the operator U must be *unitary*. A matrix U is said to be unitary if its inverse is equivalent to its conjugate transpose (which is also unitary), i.e., $U^{-1} = U^\dagger$. This brings us the next requirement for quantum operations; aside from measurement, they must be *reversible*. We can always "undo" the transformation performed by U by applying U^{-1}. This is in contrast with classical model, in which operations are not necessarily reversible, with one example being erasing data from memory.

Quantum circuits are constructed using a sequence of *single-qubit gates* and *two-qubit gates*, which are unitary operators whose only nontrivial action is limited to single- or two-qubit subsystems. Perhaps the most well-known single-qubit gate is the *Hadamard gate*:

$$H = \frac{1}{\sqrt{2}} \begin{pmatrix} 1 & 1 \\ 1 & -1 \end{pmatrix},$$

which acts on the states $|0\rangle$ and $|1\rangle$ as follows:

$$H|0\rangle = \frac{1}{\sqrt{2}} \left(|0\rangle + |1\rangle \right), \quad H|1\rangle = \frac{1}{\sqrt{2}} \left(|0\rangle - |1\rangle \right).$$

In other words, the Hadamard gate transforms a basis state into a superposition. One can also easily verify that the Hadamard transform is its own inverse.

The *Pauli* gates are also foundational single-qubit gates, and are defined as follows:

$$I = \begin{pmatrix} 1 & 0 \\ 0 & 1 \end{pmatrix}, \quad X = \begin{pmatrix} 0 & 1 \\ 1 & 0 \end{pmatrix}, \quad Y = \begin{pmatrix} 0 & -i \\ i & 0 \end{pmatrix}, \quad Z = \begin{pmatrix} 1 & 0 \\ 0 & -1 \end{pmatrix}.$$

Here, I is the usual identity which renders the state unchanged, X is the quantum analogue of the classical NOT gate (i.e., it is a qubit flip operation), while Z is a phase flip operation, leaving $|0\rangle$ unchanged and mapping $|1\rangle$ to $-|1\rangle$.

While the Hadamard and Pauli gates act on a single qubit, we can apply them to a larger system by taking the tensor product with the identity on the rest of the system.

2.5.3 Measurement

Whereas with classical computers we can directly read the state of the bits, information is obscured in the quantum case. We cannot simply observe whatever superposition our state $|\phi\rangle$ might be in, the best we can hope to do is gain some classical information from $|\phi\rangle$ by applying a *measurement gate* to it. However, measurements are destructive; after measurement the state

$$|\phi\rangle = \sum_{i=0}^{N-1} \alpha_i |i\rangle,$$

collapses to a single classical state $|i\rangle$, and in particular we see state $|i\rangle$ with probability $|\alpha_i|^2$. In other words, if we apply a measurement gate to $|\phi\rangle$, and observe the classical state i, then any other information on $|\phi\rangle$ has been destroyed, and repeating the experiement will yield the same outcome.

2.5.4 Mixed states

A key component of the algorithms we will study in **Part III** are *mixed quantum states*, which are probabilistic mixtures of pure states. Mixed quantum states are represented using *density operators*; letting $(p)_{i=1}^{k}$ be a probability distribution over pure states $|\phi_i\rangle \in (\mathbb{C}^2)^{\log(n)}$, we have

$$\rho = \sum_{i=1}^{k} p_i |\phi_i\rangle\langle\phi_i| \in \mathbb{C}^{n \times n}.$$

That is, $\log(n)$-qubit mixed states/n-dimensional density operators are trace-normalized positive semidefinite matrices. Note however, while any density matrix can be decomposed into pure states, the factorization is not unique in general.

2.5.5 Quantum accessible data structures

In [KP16], Kerenidis and Prakash introduced a data structure, that when stored in Quantum Random Access Memory (QRAM), allows one to perform queries over a superposition of addresses.

Let $v_j \in \mathbb{R}^{2^q}$ be a classical vector, and a consider the quantum state $\sum_{j=0}^{2^q-1} \alpha_j |j\rangle$. If v_j is stored in a QRAM, it is assumed that the mapping

$$\sum_{j=0}^{2^q-1} \alpha_j |j\rangle \otimes |0\rangle \rightarrow \sum_{j=0}^{2^q-1} (\alpha_j |j\rangle \otimes |v_j\rangle), \tag{2.6}$$

can be performed in essentially constant time (some dependence on the number of bits of each entry could be allowed, but it is convenient to ignore such details for the simplified exposition presented in this entry). Here, j is the address in superposition with the amplitudes α_j proportional to the elements v_j. There are different ways (see e.g., [GLM08, LKS18]) for one to implement the operation in (2.6), so the user can optimize this choice for their particular setting.

The next result summarizes key aspects of the QRAM we consider.

Theorem 2.3 ([KP20]). *Let $M \in \mathbb{R}^{m \times n}$ be a matrix with $M_{ij} \in \mathbb{R}$ being the entry of the i-th row and the j-th column. If w is the number of non-zero entries of M then, there exists a data structure of size $\mathcal{O}\left(w \log^2(mn)\right)$, that, given the entries (i, j, M_{ij}) in an arbitrary order, stores them such that time taken to store each entry of M is $\mathcal{O}\left(\log(mn)\right)$. Once this data structure has been initiated with all non-zero entries of M, there exists a quantum algorithm that can perform the following maps with ϵ-precision in $\mathcal{O}(\operatorname{polylog}(mn/\epsilon))$ time:*

$$\hat{U} : |i\rangle|0\rangle \mapsto |i\rangle \frac{1}{\|M_{i,\cdot}\|} \sum_{j=1}^{n} M_{ij} |j\rangle = |i, M_{i,\cdot}\rangle,$$

$$\hat{V} : |0\rangle|j\rangle \mapsto \frac{1}{\|M\|_F} \sum_{i=1}^{m} \|M_{i,\cdot}\| |i\rangle |j\rangle = |\widehat{M}, j\rangle,$$

where $|M_{i,\cdot}\rangle$ is the normalized quantum state corresponding to the i-th row of M and $|\widehat{M}\rangle$ is a *normalized quantum state such that $\langle i | \widehat{M} \rangle = \| M_{i,\cdot} \|$, i.e., the norm of the i-th row of A.*

In particular, given a vector $v \in \mathbb{R}^{n \times 1}$ stored in this data structure, we an generate a ϵ-approximation of the superposition $\frac{1}{\|v\|} \sum_i v_i |i\rangle$ with complexity $\mathcal{O}(\mathrm{polylog}(n/\epsilon))$.

For an extensive discussion on the details of the QRAM described here, we refer the reader to [KP16, KP20].

2.5.6 Input models and subroutines

We provide analyses that consider two input models: the *sparse-access model*, and the *quantum operator model*. We do not consider the *quantum state model* [BKL$^+$19, vAG19], as the quantum operator model subsumes both the quantum state model and the sparse-access model.

Sparse-access model

In the *sparse-access model*, the input matrix $A \in \mathbb{R}^{n \times n}$ is assumed to be s-row sparse for some known bound $s \in [n]$. In other words, A has at most s nonzero entries per row. The sparse-access model is closely related to the classical notion, in that we assume access to an oracle O_{sparse}, which upon being queried with input (i, j) returns the index of the j-th nonzero entry of the i-th row of A by calculating the index function:

$$\mathrm{index} : [n] \times [s] \to [n].$$

That is, for $i \in [n]$ and $j \in [s]$, O_{sparse} computes the position in place:

$$O_{\mathrm{sparse}} |i, j\rangle = |i, \mathrm{index}(i, j)\rangle.$$

Additionally, we assume access to another oracle that returns a bitstring representation of the individual entries for every $i, j \in [n]$:

$$O_A |i, j, z\rangle = |i, j, z \oplus (A_{ij})\rangle.$$

Quantum operator model

The *quantum operator model* assumes access to an oracle O_U which acts in the following manner:

$$O_U |\psi\rangle = U |\psi\rangle.$$

Under this framework, we *block-encode* a matrix $A \in \mathbb{R}^{n \times n}$ as a unitary U which satisfies:

$$U = \begin{pmatrix} \frac{A}{\alpha} & \cdot \\ \cdot & \cdot \end{pmatrix},$$

where α is a subnormalization factor chosen to ensure $\|U\|\leq 1$. We provide a detailed overview of block-encodings in Chapter 3.

2.5.7 Quantum state tomography

Quantum state tomography is a procedure for classically estimating quantum states. The current state of the art is the algorithm from van Apeldoorn et al. [vACGN22], whose (near optimal) algorithm built upon recent advances in quantum gradient estimation [GAW19]. Letting T_U be the time to implement the unitary that prepares our state of interest $|\psi\rangle$, we can apply the algorithm in [vACGN22] to obtain a ϵ-precise (in the ℓ_2-norm) classical estimate of an n-dimensional quantum state $|\psi\rangle$ in time

$$\widetilde{\mathcal{O}}_{n,\frac{1}{\epsilon},\frac{1}{\delta}}\left(T_U\frac{n}{\epsilon}+\frac{n^2}{\epsilon}\right),$$

where $1-\delta$ is the probability of success. This is formalized in the following result.

Theorem 2.4 ([vACGN22]). *Let $|\psi\rangle = \sum_{j=0}^{n-1} v_j|j\rangle$ be a quantum state, $v \in \mathbb{C}^n$ the vector with elements v_j, and $U|0\rangle = |\psi\rangle$. There is a quantum algorithm that, with probability at least $1-\delta$, outputs $\tilde{v} \in \mathbb{R}^n$ such that $\|\Re(v)-\tilde{v}\|_2 \leq \epsilon$ using $\mathcal{O}(\frac{n}{\epsilon}\ln\frac{n}{\delta})$ applications of U and $\mathcal{O}(\frac{n^{1.5}}{\epsilon}\log\frac{n}{\epsilon}\log\frac{1}{\delta})$ additional gates; the total gate complexity of the algorithm is therefore $\widetilde{\mathcal{O}}_{n,\frac{1}{\epsilon},\frac{1}{\delta}}\left(T_U\frac{n}{\epsilon}+\frac{n^2}{\epsilon}\right)$.*

We point out that the base algorithm in [vACGN22] produces ℓ_∞-norm estimates, and the ℓ_2-norm estimate is simply a result of accounting for norm-equivalence in the choice of the error tolerances, i.e, setting the ℓ_∞-norm error to $\epsilon/\sqrt{n}$. It is shown in [vACGN22] that the number of applications of U cannot be reduced in general. It is however possible to remove the quadratic dependence on the dimension provided access to QRAM; the dependence follows from the use of a number of indexed SWAP gates, which can naturally carried out as a "QRAM operation" provided access to a sufficiently large QRAM to store the vectors. This is formalized in the next result from [ANTZ21], which is adapted from [vACGN22, Prop. 22].

Theorem 2.5 (Theorem 2 in [ANTZ21]). *Let $|\psi\rangle = \sum_{j=0}^{n-1} v_j|j\rangle$ be a quantum state, $v \in \mathbb{C}^n$ the vector with elements v_j, and $U|0\rangle = |\psi\rangle$. There is a quantum algorithm that, with probability at least $1-\delta$, outputs $\tilde{v} \in \mathbb{R}^n$ such that $\|\Re(v)-\tilde{v}\|_2 \leq \epsilon$ using $\mathcal{O}(\frac{n}{\epsilon}\log\frac{n}{\delta})$ applications of U, $\widetilde{\mathcal{O}}_{n,\frac{1}{\delta},\frac{1}{\epsilon}}(\frac{n}{\epsilon})$ indexed-SWAP gates acting on n bits, and $\widetilde{\mathcal{O}}_{n,\frac{1}{\delta},\frac{1}{\epsilon}}(\frac{n}{\epsilon})$ additional gates. If we have access to a classical-write, quantum-read QRAM of size $\widetilde{\mathcal{O}}_{n,\frac{1}{\delta},\frac{1}{\epsilon}}(2^{n/\epsilon})$, we do not need the indexed-SWAP gates.*

Proof. This follows from [vACGN22, Prop. 22], setting the ℓ_∞-norm error to $\epsilon/\sqrt{n}$. For the gate complexity, [vACGN22] uses a "QRAM-like" indexed-SWAP gate, which is used to efficiently implement the mapping from a binary description of a vector x to its amplitude encoding $|x\rangle$. It is straightforward to note that this mapping can also be stored in a classical-write, quantum-read QRAM: as the vectors $x \in [-1,1]^n$ necessary for the construction are known in advance, and each coordinate requires precision $\mathcal{O}(1/\epsilon)$, we can employ the usual QRAM data structure for each of these $\mathcal{O}(2^{n/\epsilon})$ vectors in QRAM. This operation needs to be performed only once and it only depends on the dimension and the precision. Alternatively, the operation can be performed with a $\widetilde{\mathcal{O}}_{n,\frac{1}{\delta},\frac{1}{\epsilon}}\left(\frac{n^2}{\epsilon}\right)$ quantum-writable QRAM (the data structure can be computed in superposition and written onto the QRAM). $\qquad\square$

Part I

Quantum Linear Algebra

Chapter 3

Quantum Linear Algebra using Block-Encodings

This chapter is based on the book chapter [AN23].

3.1   Introduction

The underlying paradigm of quantum computers fundamentally differs from the classical notion; in the quantum picture we apply *unitary operators*, and use them to evolve *quantum states* as opposed to performing operations on matrices and vectors. Another key consideration is how the input is presented. Conveniently, the *quantum operator input model* [LC19, GSLW19], provides a foundation for both input and computation that can be largely understood if one has a working knowledge of conventional linear algebra. The key mechanism of the quantum operator input model is to encode classical data into unitary operators known as *block-encodings*. A block-encoding of an $n \times n$ matrix A is a unitary which stores A in their top-left block:

$$U = \begin{pmatrix} A/\alpha & \cdot \\ \cdot & \cdot \end{pmatrix},$$

where $\alpha \geq \|A\|$ is a factor used to normalize U with respect to the operator norm $\|\cdot\|$. The only role of the other blocks of U is to ensure that U is indeed unitary, and usually these entries are all 0 (provided that we properly set α).

Kerenidis and Prakash [KP16] showed that when our matrix A is stored in a quantum data structure, one could implement a ξ-approximate block-encoding of A with complexity of the order polylog(n/ξ). Perhaps remarkably, it has also been shown (see, e.g. [CGJ19, Gil19, GSLW19]) that one can implement operations on unitaries that correspond to operations such as matrix products, as well as taking powers of matrices in time that is polylogarithmic in the problem dimension, under some conditions. These techniques and results can be extended to matrices that are given in the sparse-access input model, which closely follows the classical notion of sparse-access to matrices [Gil19]. In fact, the quantum operator input model generalizes many possible input models, providing a unifying framework for analyzing the cost of quantum optimization algorithms [ANTZ21, Gil19, GSLW19, KP20, vAG19], and will be utilized in every algorithm studied in this thesis.

The aim of this chapter is to provide an overview on these techniques that build the foundation for the algorithms we study in later chapters.

3.2   Implementing block-encoded matrices

We now formally define the block-encoding of a matrix A. This definition makes use of three quantities *(i)* α; the subnormalization factor we saw from earlier, *(ii)* a; the number of ancilla qubits, and *(iii)* ξ; the error to which we implement U.

Definition 3.1 (Block-encoding). *Suppose that A is an w-qubit operator (i.e., $A \in \mathbb{C}^{2^w \times 2^w}$), $\alpha, \xi \in \mathbb{R}$, and $a \in \mathbb{N}$. We say that the unitary matrix $U \in \mathbb{C}^{2^{w+a} \times 2^{w+a}}$ is an (α, a, ξ)-block-encoding of A if:*

$$\|A - \alpha(\langle 0|^{\otimes a} \otimes I^{\otimes w})U(|0\rangle^{\otimes a} \otimes I^{\otimes w})\| \leq \xi.$$

Remark 3.1. *The above definition simply means that the top-left block of U of size $2^w \times 2^w$ is close to A/α in the spectral norm: the expression $(\langle 0|^{\otimes a} \otimes I^{\otimes w})U(|0\rangle^{\otimes a} \otimes I^{\otimes w})$ is just the aforementioned submatrix.*

By definition, a unitary matrix is a $(1, 0, 0)$-block encoding of itself, and is thus referred to as a *trivial block-encoding*.

In order to construct a block-encoding of a matrix A, we follow the construction presented in [CGJ19]. Suppose we want to construct a matrix A, and that we have a way of constructing the quantum states $|\psi_i\rangle = \sum_j \frac{A_{ij}}{\|A_i\|}|i, j\rangle$, where A_i is the i-th row of A, and $|\phi_j\rangle = \sum_i \frac{\|A_i\|}{\|A\|_F}|i, j\rangle$. Then we construct:

$$U_R^\dagger U_L = \begin{pmatrix} - & \langle\psi_1| & - & \cdot \\ & \vdots & & \\ - & \langle\psi_n| & - & \cdot \\ & \cdot & & \end{pmatrix} \begin{pmatrix} | & & | & \\ |\phi_1\rangle & \cdots & |\phi_n\rangle & \cdot \\ | & & | & \\ & \cdot & & \cdot \end{pmatrix} = \begin{pmatrix} [\langle\psi_i||\phi_j\rangle]_{i,j\in[n]} & \cdot \\ & \cdot \end{pmatrix} = \begin{pmatrix} \frac{A}{\|A\|_F} & \cdot \\ \cdot & \cdot \end{pmatrix},$$

where we used the fact that

$$\langle\psi_i|\phi_j\rangle = \frac{A_{ij}}{\|A\|_F}.$$

Simply put, $U_R^\dagger U_L$ is a block-encoding of A with normalization factor $\alpha = \|A\|_F$.

The construction we have just outlined assumes that have a procedure to implement U_R, U_L. Fortunately, the recipe for these unitaries is straightforward, starting from what are known as *controlled operations* to prepare the states $|\psi_i\rangle, |\phi_j\rangle$, i.e., operations of the form

$$|i\rangle|0\rangle \to |i\rangle|\psi_i\rangle$$

$$|j\rangle|0\rangle \to |j\rangle|\phi_j\rangle.$$

If we allow pre-processing to create certain data structures that can be stored in QRAM, we can improve the efficiency of this construction.

Proposition 3.1 (Lemma 47 in [GSLW19]). *(Block-encoding of Gram matrices) Let U_L and U_R be state preparation unitaries acting on $s + d$ preparing the vectors $\{|\psi_i\rangle : i \in [2^s] - 1\}$, $\{|\phi_j\rangle : j \in [2^s] - 1\}$ as:*

$$U_L : |0\rangle^{\otimes d} \otimes |i\rangle \to |\psi_i\rangle$$

$$U_R : |0\rangle^{\otimes d} \otimes |j\rangle \to |\phi_j\rangle.$$

Then $U = U_L^\dagger U_R$ is a $(1, d, 0)$-block-encoding of the Gram matrix A defined as $A_{ij} = \langle \psi_i | \phi_j \rangle$.

The result of Proposition 3.1 provides an avenue for block-encoding matrices to which we have sparse-oracle access. Assuming that the matrix $A \in \mathbb{C}^{m \times n}$ we wish to block-encode has s_r-sparse rows and s_c-sparse columns, suppose that we have oracle access to the elements of A:

$$O_A : |i\rangle|j\rangle|0\rangle^{\otimes p} \mapsto |i\rangle|j\rangle|A_{ij}\rangle, \quad \forall i \in [m], j \in [n].$$

If we further assume access to oracles O_R and O_R that are capable of querying the indices of non-zero elements of each row and column, we can implement an efficient algorithmic description of the sparsity pattern and element values of A. The following result from [GSLW19] formalizes how one can efficiently implement block-encodings of sparse-access matrices.

Lemma 3.2 (Lemma 48 in [GSLW19]). *Let $A \in \mathbb{C}^{2^w \times 2^w}$ be a matrix that is s_r-row-sparse and s_c-column-sparse, and each element of A has absolute value at most 1. Suppose that we have access to the following sparse-access oracles acting on two $(w + 1)$ qubit registers:*

$$O_r : |i\rangle|k\rangle \mapsto |i\rangle|r_{ik}\rangle \quad \forall i \in [2^w] - 1, k \in [s_r], \ and$$

$$O_c : |\ell\rangle|j\rangle \mapsto |c_{\ell j}\rangle|j\rangle \quad \forall \ell \in [s_c], j \in [2^w] - 1, \ where$$

r_{ij} is the index for the j-th non-zero entry of the i-th row of A, or if there are less than i non-zero entries, then it is $j + 2^w$, and similarly c_{ij} is the index for the i-th non-zero entry of the j-th column of A, or if there are less than j non-zero entries, then it is $i + 2^w$. Additionally, assume that we have access to an oracle O_A that returns the entries of A in a binary description:

$$O_A : |i\rangle|j\rangle|0\rangle^{\otimes p} \mapsto |i\rangle|j\rangle|a_{ij}\rangle, \quad \forall i, j \in [2^w] - 1,$$

where a_{ij} is a p-bit binary description of the ij-matrix element of A. Then, we can implement a $(\sqrt{s_r s_c}, w + 3, \xi)$-block-encoding of A with a single use of O_r, O_c and two uses of O_A, and additionally using $\mathcal{O}\left(w + \log^{2.5}\left(\frac{s_r s_c}{\xi}\right)\right)$ one and two qubit gates while using $\mathcal{O}\left(p + \log^{2.5}\left(\frac{s_r s_c}{\xi}\right)\right)$ ancilla qubits.

We now turn our attention to block-encoding matrices that are stored in the QRAM data structure that we described in Section 2.5.5. With quantum read accesses to A, we can prepare a a block-encoding of A, and subsequently perform many operations on the resulting unitary in time that is merely polylogarithmic in the size of the matrix. In constructing a block-encoding of a matrix stored in QRAM, it could be at time useful to choose a norm other than $\|\cdot\|_F$ for the subnormalization factor. Specifically, one can use the "q-norm", which is typically defined in the literature as

$$\mu_q(A) := \sqrt{\max_i \|A_{i,\cdot}\|_q^q \cdot \max_i \|A_{i,\cdot}^\top\|_{2-q}^{2-q}},$$

for $q \in [0, 2]$.

We now formalize the construction of block-encodings of matrices stored in QRAM.

Proposition 3.3 (Lemma 50 in [GSLW19]). *Let $A \in \mathbb{C}^{m \times m}$ with $m = 2^w$ and $\xi > 0$.*

(i) *Define $A^{\{q\}}$ to be the matrix with elements $(A^{\{q\}})_{ij} = \sqrt{A_{ij}^q}$. If $A^{\{q\}}$ and $(A^{\{2-q\}})^\dagger$ are both stored in QRAM data structures, then there exist unitaries U_R and U_L that can be implemented in time $\mathcal{O}(\mathrm{poly}(w \log \frac{1}{\xi}))$ and such that $U_R^\dagger U_L$ is a $(\mu_q(A), w + 2, \xi)$-block-encoding of A.*

(ii) *If A is stored in a QRAM data structure, then there exist unitaries U_R and U_L that can be implemented in time $\mathcal{O}(\mathrm{poly}(w \log \frac{1}{\xi}))$ and such that $U_R^\dagger U_L$ is an $(\|A\|_F, w + 2, \xi)$-block-encoding of A.*

A detailed analysis of the resources required to block-encode classical matrices can be found in [CDS+22], who study two distinct models of QRAM.

Having understood the construction of block-encodings, we can now discuss linear algebra in this framework.

3.3 Basic operations with block-encodings

We begin with a discussion on how to perform basic matrix arithmetic, such as products and linear combinations using a quantum computer.

3.3.1 Matrix-vector products

Suppose we want to multiply A by some vector x, and assume that we have access to a quantum state $|x\rangle$ encoding x. In the quantum model of computation, computing the matrix vector product Ax is equivalent to updating the state $|x\rangle$ in the following manner:

$$|x\rangle \to \frac{A|x\rangle}{\|A|x\rangle\|}.$$

Hence, if U is an (α, a, ξ)-block encoding of A up to total error ξ, we have

$$U|u\rangle|0\rangle^{\otimes a} \approx \begin{pmatrix} \frac{A}{\alpha} & \cdot \\ \cdot & \cdot \end{pmatrix} \begin{pmatrix} |u\rangle \\ |0\rangle \end{pmatrix} = \frac{1}{\alpha}A|u\rangle|0\rangle^{\otimes a} + |\cdot\rangle.$$

By performing approximately $\frac{\alpha}{\|A|u\rangle\|}$ rounds of amplitude amplification, we obtain $\frac{A|u\rangle}{\|A|u\rangle\|}$ with high probability, see e.g., [CGJ19].

3.3.2 Linear combinations of matrices

Given access to block encodings of m matrices $A_1, \ldots, A_m$, one can also implement a linear combination of them, i.e., we can implement a block-encoding of

$$\sum_{j=1}^{m} y_j A_j,$$

where, the y_j terms are some appropriately chosen weights. This technique was first proposed by Berry et al. [BCC$^+$15], and later used by Childs, Kothari and Somma [CKS17] to improve the dependence of QLSAs with respect to the precision from polynomial to polylogarithmic.

Definition 3.2. *(State preparation pair) Let $y \in \mathbb{C}^m$ with $\|y\|_1 \leq \beta$. The pair of unitaries P_L, P_R are a (β, b, ϵ)-state-preparation pair if $P_L|0\rangle^{\otimes b} = \sum_{j=0}^{2^b-1} c_j|j\rangle$, $P_R|0\rangle^{\otimes b} = \sum_{j=0}^{2^b-1} d_j|j\rangle$ with the property that $\sum_{j=0}^{m-1} |\beta(c_j^* d_j) - y_j| \leq \epsilon$ and for all $j = m, \ldots, 2^b - 1$ we have $c_j^* d_j = 0$.*

Adequately setting the weights allows us to perform addition and subtraction, and hence, implement a block-encoding of a linear combination of block-encodings.

Proposition 3.4 (Lemma 52 in [GSLW19]). *(Linear combination of block-encoded matrices, with weights given by the coefficients of a basis state) Let $A = \sum_{j=1}^{m} y_j A_j$ be an s-qubit operator, where A_j are matrices. Suppose P_L, P_R is a (β, b, ξ_1)-state-preparation pair for y, $W = \sum_{j=0}^{m-1} |j\rangle\langle j| \otimes U_j + ((I - \sum_{j=0}^{m-1} |j\rangle\langle j|) \otimes I_a \otimes I_s)$ is an $(s+a+b)$-qubit unitary with the property that U_j is an (α, a, ξ_2)-block-encoding of A_j. Then we can implement a $(\alpha\beta, a + b, \alpha\xi_1 + \alpha\beta\xi_2)$-block-encoding of A with a single use of W, P_R and $P_L^\dagger$.*

Using Proposition 3.4, one can devise a prepare block-matrices, provided that each of the blocks is stored in an efficient data structure. The idea reduces to block-encoding each of the block-factors by block-encoding a matrix that stores the block in the same position as it appears in A, with all other elements being zero and simply summing the resulting rc-many matrices, which yields A.

Proposition 3.5. *Let*

$$A = \begin{pmatrix} M_{11} & \dots & M_{1c} \\ \vdots & \ddots & \vdots \\ M_{r1} & \dots & M_{rc} \end{pmatrix} \in \mathbb{R}^{n \times n},$$

where for simplicity r, c are powers of 2, and each M_{ij} is a matrix, of appropriate dimension, that is stored in a QRAM data structure. Suppose further the row norms and column norms of each matrix M_{ij} are known. Then we can construct a $(rc\|A\|_F, \mathcal{O}(\log n), \xi)$-block-encoding of A in time $\mathcal{O}(\text{poly}(rc, \log n, \log \frac{1}{\xi}))$.

Proof. Let $A_1, \dots, A_m$ denote the $m = rc$ matrices in $\mathbb{R}^{n \times n}$ that each store one of the M_{ij} matrices in the same position that it appears in A, but with all other entries being 0. Then, A can be written as a linear combination of the m many matrices A_k as $A = \sum_{k=1}^m y_k A_k$, where $y_k = 1$ for all k. We normalize all matrices using the Frobenius norm of A, to ensure that all spectral norms are ≤ 1.

Notice that if a matrix M_{ij} is stored in QRAM, we can construct the block-encoding of the corresponding A_k using Proposition 3.3, with straightforward modifications of the underlying algorithm, because the elements and norms of A_k can be efficiently computed from those of M_{ij}. Thus, a $(\|A\|_F, \log n + 2, \xi')$-block-encoding of A_k can be efficiently constructed, for any $\xi' > 0$.

Let P_L, P_R be a $(m, \log m, 0)$-state-preparation pair for y, which can be constructed by simply taking $P_L = P_R = H^{\otimes \log m}$ where H is the Hadamard gate. We then use Proposition 3.4, where W can be obtained by adding a control qubit to the circuits used to construct the block-encoding of each A_k, and the parameter ξ_2 is chosen to be $\xi/(m\|A\|_F)$. This yields a $(m\|A\|_F, \log n + \log m, \xi)$-block-encoding of A with a single use of W, P_R and $P_L^\dagger$; since P_R and $P_L^\dagger$ only require $\mathcal{O}(\log m)$ gates each, and W takes $\mathcal{O}(\text{poly}(m, \log n, \log \frac{1}{\xi}))$ time, the proof is complete. $\square$

3.3.3 Matrix multiplication

We now turn our attention to matrix multiplication. Suppose that U_A is an (α_1, a_1, ξ_A)-block-encoding of an w-qubit operator A, and that U_B is a (α_2, a_2, ξ_B)-block-encoding of a w-qubit operator B. Then we have

$$(I_{a_2} \otimes U_A)(I_{a_1} \otimes U_B) = \begin{pmatrix} \frac{B}{\alpha_2} & \cdot \\ \cdot & \cdot \end{pmatrix} \begin{pmatrix} \frac{A}{\alpha_1} & \cdot \\ \cdot & \cdot \end{pmatrix} = \begin{pmatrix} \frac{AB}{\alpha_1 \alpha_2} & \cdot \\ \cdot & \cdot \end{pmatrix},$$

which yields an $(\alpha_1 \alpha_2, a_1 + a_2, \alpha_1 \xi_B + \alpha_2 \xi_A)$-block-encoding of AB. In the above statement $(I_{a_2} \otimes U_A)(I_{a_1} \otimes U_B)$ indicates the identity being applied to each operator's auxiliary qubits.

The next two propositions concern the construction the product of two block-encoded matrices, which can be accomplished with at cost that is polylogarithmic in the size of the matrices, and the resulting error is bounded by the sum of the errors of our block-encodings for A and B.

Proposition 3.6 (Lemma 4 [CGJ19]). *(Product of two block-encoded matrices) Given an (α_1, a_1, ξ_A)-*

block-encoding U_A of an s-qubit operator A, and an (α_2, a_2, ξ_B)-block-encoding U_B of an s-qubit operator B, $(I_{a_2} \otimes U_A)(I_{a_1} \otimes U_B)$ is an $(\alpha_1\alpha_2, a_1 + a_2, \alpha_1\xi_B + \alpha_2\xi_A)$-block-encoding of AB.

We may be able to slightly improve the construction in situations where the normalization factor of the block encoding of the product is fixed, as opposed to depending on the input. This is captured by the next result.

Proposition 3.7 (Lemma 5 in [CGJ19]). *(Product of preamplified block-encoded matrices) Suppose A and B have sizes such that AB is a a valid matrix product, and $\|A\|, \|B\| \leq 1$. Given an (α_1, a_1, ξ_1)-block-encoding U_A of an s-qubit operator A, and an (α_2, a_2, ξ_2)-block-encoding U_B of an w-qubit operator B, with $\alpha_1 \geq 1, \alpha_2 \geq 1$, then we can implement a $(2, a_1 + a_2 + 2, \sqrt{2}(\xi_1 + \xi_2 + \gamma))$-block-encoding of AB in time $\mathcal{O}((\alpha_1(T_{U_A} + a_2) + \alpha_2(T_{U_B} + a_1)) \log\frac{1}{\gamma})$, where T_{U_A} and T_{U_B} are the implementation times for U_A and U_B, respectively.*

These resulted can be generalized to products of m block-encoded matrices, and it turns out that the resulting level error is at worst quadratic in the number of matrices.

Proposition 3.8 (Corollary 55 in [GSLW19]). *(Product of multiple block-encoded unitaries) Given unitaries U^j, each of which is a $(1, a, \epsilon)$-block-encoding of a unitary W^j for $j \in [K]$, the unitary $\prod_{j=1}^{m} U^j$ is a $(1, a, 4K^2\epsilon)$-block encoding of $\prod_{j=1}^{m} W^j$.*

Remark 3.2. *The above proposition is for unitaries, not general matrices.*

3.3.4 Kronecker and Hadamard products

The very definition of block-encoding suggests that one can naturally construct the Kronecker product of block-encodings. One can block-encode $A \otimes B$ by simply taking the tensor product of block-encodings for A and B (though it is important that we keep the ancilla qubits separate).

Proposition 3.9. *(Implementing the Kronecker product of block-encoded matrices) Suppose that U_A is an (α_1, a_1, ξ_A)-block-encoding of $A \in \mathbb{R}^{n \times n}$, and U_B is an (α_2, a_2, ξ_B)-block-encoding of $B \in \mathbb{R}^{n \times n}$. Then, taking the tensor product of U_A and U_B, we obtain a $(\alpha_1\alpha_2, a_1 + a_2, \xi_A + \xi_B)$-block-encoding of $A \otimes B$.*

Some optimization algorithms, such as interior point methods, make use of what is known as the *symmetric Kronecker product*, as it provides a useful way to construct the *Newton linear system*. It should be straightforward from Definition 2.2 to see how the results in Propositions 3.4, 3.6 and 3.9 allow us to block-encode the symmetric Kronecker product of two matrices. At a high level, the process requires five steps:

1. Block-encode A, B, V and $V^\top$.

2. Use the block-encodings of A and B to construct block-encodings of $A \otimes B$ and $B \otimes A$.

3. Take the linear combination of our block-encodings of $A \otimes B$ and $B \otimes A$ to implement a block-encoding of $A \otimes B + B \otimes A$.

4. Take the product of block-encodings of V and $(A \otimes B + B \otimes A)$ to block-encode $V(A \otimes B + B \otimes A)$.

5. Take the product of block-encodings of $V(A \otimes B + B \otimes A)$ and $V^\top$ to block-encode $V(A \otimes B + B \otimes A)V^\top$.

The next result formalizes this notion.

Proposition 3.10. *(Implementing the symmetric Kronecker product of block-encoded matrices)* *Suppose that U_A is an (α_1, a_1, ξ_A)-block-encoding of $A \in \mathbb{R}^{n \times n}$, and U_B is a (α_2, a_2, ξ_B)-block-encoding of $B \in \mathbb{R}^{n \times n}$. Then, a $(2(\alpha_1\alpha_2)^2, 2(a_1 + a_2) + 2\log(n) + 6, 7\alpha_1\alpha_2(\xi_A + \xi_B))$-block-encoding of $A \otimes_s B$ can be constructed using two applications of U_A and U_B each, and $\tilde{\mathcal{O}}_n(1)$ additional gates.*

Proof. Recall that

$$A \otimes_s B = \frac{1}{2}V \left(A \otimes B + B \otimes A\right) V^\top,$$

where V is the $\frac{n(n+1)}{2} \times n^2$ matrix defined in Definition 2.2. Observe that by definition, V is 1-column-sparse and 2-row-sparse so $s_c s_r = 1 \cdot 2 = 2$.

Using the description of V, we can construct the sparse-access oracles O_r and O_c as defined in Lemma 3.2 (which act on two $(\log n^2 + 1)$ qubit registers). Additionally, from the definition of V we can construct an oracle O_V that returns the entries of V in a binary description:

$$O_V : |i\rangle|j\rangle|0\rangle^{\otimes p} \mapsto |i\rangle|j\rangle|v_{ij}\rangle, \quad \forall i, j \in [2^{\log n^2}] - 1,$$

where v_{ij} is a p-bit binary description of the ij-matrix element of V. By Lemma 3.2 we can efficiently implement a $(\sqrt{2}, 3 + \log(n), \xi_V)$-block-encoding of V with a single use of O_r, O_c and two uses of O_V. We can construct a $(\sqrt{2}, 3 + \log(n), \xi_V)$-block-encoding of $V^\top$ in the same manner.

Given that U_A is an (α_1, a_1, ξ_A)-block-encoding of $A \in \mathbb{R}^{n \times n}$, and U_B is a (α_2, a_2, ξ_B)-block-encoding of $B \in \mathbb{R}^{n \times n}$, we can apply Proposition 3.9 to construct an $(\alpha_1\alpha_2, a_1 + a_2, \xi_A + \xi_B)$-block-encoding of $A \otimes B$ using one application of U_A and U_B and no additional gates. Clearly, an $(\alpha_1\alpha_2, a_1 + a_2, \xi_A + \xi_B)$-block-encoding for $B \otimes A$ can be constructed in exactly the same way.

From here we use a linear combination of our block-encodings of $A \otimes B$ and $B \otimes A$ to block-encode $A \otimes B + B \otimes A$. Applying Proposition 3.4 yields an $((\alpha_1\alpha_2)^2, 2(a_1 + a_2), 2\alpha_1\alpha_2(\xi_A + \xi_B))$-block-encoding of $A \otimes B + B \otimes A$.

Using our block-encodings of V and $A \otimes B + B \otimes A$, we apply Proposition 3.6 with

$$\xi_V = \frac{\xi_A + \xi_B}{\alpha_1\alpha_2}$$

to implement a $(\sqrt{2}(\alpha_1\alpha_2)^2, 2(a_1 + a_2) + \log(n) + 3, 3\sqrt{2}\alpha_1\alpha_2(\xi_A + \xi_B))$-block-encoding of

$$V\left(A \otimes B + B \otimes A\right).$$

A final application of Proposition 3.6 with

$$\xi_V = \frac{\xi_A + \xi_B}{\alpha_1\alpha_2}$$

yields a $(2(\alpha_1\alpha_2)^2, 2(a_1 + a_2) + 2\log(n) + 6, 7\alpha_1\alpha_2(\xi_A + \xi_B))$-block-encoding of

$$\frac{1}{2}V\left(A \otimes B + B \otimes A\right)V^\top = A \otimes_s B.$$

The stated complexity result follows upon noting that the steps required to construct the unitary

$$U_{A\otimes_s B} = U_V\left(U_{A\otimes B} + U_{B\otimes A}\right)U_{V^\top},$$

requires one application of $U_{A\otimes B}$, one application of $U_{B\otimes A}$ and one application of each of the other matrices. In turn, this amounts to 2 applications of U_A and U_B each, plus the $\tilde{\mathcal{O}}_n(1)$ gate cost of the matrices U_V and $U_{V^\top}$, and the proof is complete. $\qquad\square$

We can also implement the *Hadamard* or element-wise product of two matrices $A \circ B$, which is a simply the principle submatrix of the Kronecker product. The construction thus reduces to computing the Kronecker product of block-encodings, and subsequently applying transformations that can set the elements of all irrelevant rows and columns to zero, and using block-encodings of partial permutation matrices to re-order the rows and columns such that we obtain a unitary with $A \circ B$ in the top-left corner and zeros elsewhere. This is formalized in the next result.

Proposition 3.11. *(Hadamard product of block-encoded matrices) Suppose that U_A is an (α_1, a_1, ξ_A)-block-encoding of $A \in \mathbb{R}^{n\times n}$, and U_B is a (α_2, a_2, ξ_B)-block-encoding $B \in \mathbb{R}^{n\times n}$. Then, we can implement an $(\alpha_1\alpha_2, a_1 + a_2 + 8\log(n) + 12, 5(\xi_A + \xi_B))$-block-encoding of $A \circ B$ using one application of U_A and U_B, and $\tilde{\mathcal{O}}_n(1)$ additional gates.*

Proof. First, note that

$$A \circ B = (A \otimes B)[\iota_A, \iota_B],$$

where $\iota_A = \iota_B = \{1, n+2, 2n+3, \ldots, n^2\}$ are index sets of cardinality n (see, e.g., Lemma 5.1.1 in [HJ94]). Our goal is to use the index sets ι_A and ι_B along with a block encoding of $A \otimes B$ to construct a unitary which block-encodes $\mathcal{M} \in \mathbb{R}^{n^2 \times n^2}$, a matrix which contains the elements of $A \circ B$ in its upper left-most $n \times n$ block, while all other entries are 0:

$$\mathcal{M}_{ij} = \begin{cases} A_{ij} \cdot B_{ij} & \text{for } i,j = 1, \ldots, n, \\ 0 & \text{otherwise}, \end{cases}$$

i.e.,

$$\mathcal{M} = \begin{pmatrix} A \circ B & \mathbf{0}^{n \times (n^2 - n)} \\ \mathbf{0}^{(n^2 - n) \times n} & \mathbf{0}^{(n^2 - n) \times (n^2 - n)} \end{pmatrix}.$$

We will first show how one can use ι_A and ι_B to construct sparse matrices that map $A \otimes B$ to $\mathcal{M}$, and then subsequently analyze the cost of constructing the corresponding unitary block-encoding.

Consider the matrix $Z \in \mathbb{R}^{n^2 \times n^2}$, whose elements are defined as

$$Z_{ij} = \begin{cases} 1 & \text{if } i = j = (k-1)n + k, \quad k = 1, \ldots, n, \\ 0 & \text{otherwise.} \end{cases}$$

Multiplying $A \otimes B$ on the left by Z sets the rows of $A \otimes B$ which do not contain elements of $A \circ B$ to zero, and subsequently multiplying $Z(A \otimes B)$ on the right by Z will set the columns of $Z(A \otimes B)$ which do not appear in $A \circ B$ to zero. As a result, a block-encoding of $Z(A \otimes B)Z$ corresponds to block-encoding $A \otimes B$, and setting all terms not appearing in $A \circ B$ to zero:

$$[Z(A \otimes B)Z]_{ij} = \begin{cases} [A \otimes B]_{ij} & \text{if } i = (k-1)n + k \text{ and } j = (\ell-1)n + \ell \quad k, \ell = 1, \ldots, n, \\ 0 & \text{otherwise.} \end{cases}$$

Next, let $G \in \mathbb{R}^{n^2 \times n^2}$ be a matrix whose elements are defined as follows:

$$G_{ij} = \begin{cases} 1 & \text{if } i \in [n^2] \text{ and } i = j = (k-1)n + k, \quad k = 1, \ldots, n, \\ 1 & \text{if } i \in [n^2] \setminus \{1, n+2, 2n+3, \ldots, n^2\} \text{ and } j = (i-1)n + i, \\ 0 & \text{otherwise.} \end{cases}$$

We will now establish that $GZ(A \otimes B)Z)G^\top$ is precisely the matrix we seek to block-encode, by demonstrating that $G(Z(A \otimes B)Z)G^\top = \mathcal{M}$. First, observe that G is a (partial) permutation matrix: multiplying $Z(A \otimes B)Z$ on the left by G performs the necessary row-exchanges, as the elements of $G(Z(A \otimes B)Z)$ are given by

$$[G\left(Z(A \otimes B)Z\right)]_{ik} = \begin{cases} A_{ij} \cdot B_{ij} & \text{for } k = (j-1)n + j, \quad i, j = 1, \ldots, n, \\ 0 & \text{otherwise.} \end{cases}$$

On the other hand, multiplying $Z(A \otimes B)Z$ on the right by $G^\top$ performs this transformation with respect to the columns such that

$$[\left(Z(A \otimes B)Z\right)G]_{kj} = \begin{cases} A_{ij} \cdot B_{ij} & \text{for } k = (i-1)n + i, \quad i, j = 1, \ldots, n, \\ 0 & \text{otherwise.} \end{cases}$$

Hence, multiplying $G\left(Z(A \otimes B)Z\right)$ on the right by $G^\top$ conducts the column exchanges to move $A \circ B$ to the top left n-dimensional block of $Z(A \otimes B)Z$, i.e.,

$$[G\left(Z(A \otimes B)Z\right)G]_{ij} = \begin{cases} A_{ij} \cdot B_{ij} & \text{for } i,j = 1,\ldots,n, \\ 0 & \text{otherwise.} \end{cases}$$

Therefore, $G(Z(A \otimes B)Z)G^\top = \mathcal{M}$ as desired.

We now analyze the cost associated with block-encoding $\mathcal{M}$. Under the stated hypothesis, we have access to an (α_1, a_1, ξ_A)-block-encoding U_A of A, and an (α_2, a_2, ξ_B)-block-encoding U_B of B, and thus applying Proposition 3.9 we can construct an $(\alpha_1\alpha_2, a_1+a_2, \xi_A+\xi_B)$-block-encoding $U_{A \otimes B}$ of $A \otimes B$ using one application of U_A and of U_B, and no additional gates.

Using the description of Z, we can construct the sparse-access oracles O_r and O_c as defined in Lemma 3.2 (which act on two $(2\log n + 1)$ qubit registers). Additionally, from the definition of Z, we can construct an oracle O_Z, which returns the entries of Z in a binary description:

$$O_Z : |i\rangle|j\rangle|0\rangle^{\otimes p} \mapsto |i\rangle|j\rangle|z_{ij}\rangle, \quad \forall i,j \in [2^{2\log n}] - 1,$$

where z_{ij} is a p-bit binary description of the ij-matrix element of Z. Note that the circuit for the position and value of the nonzero elements of Z using $\tilde{\mathcal{O}}_n(1)$ gates because they admit an efficient description: their value is 1 and we have a compact description of their position. By construction the matrix Z is 1-row sparse and 1-column sparse, and hence an application of Lemma 3.2 with $s_r = s_c = 1$ asserts that one can construct a $(1, 2\log(n)+3, \xi_Z)$-block-encoding U_Z of Z. Given block-encodings U_Z and $U_{A \otimes B}$, we can apply Proposition 3.6 with

$$\xi_Z = \frac{\xi_A + \xi_B}{\alpha_1\alpha_2}, \quad \xi_{A \otimes B} = \xi_A + \xi_B,$$

yielding an $(\alpha_1\alpha_2, a_1 + a_2 + 2\log(n) + 3, 2(\xi_A + \xi_B))$-block-encoding of $Z(A \otimes B)$. Applying Proposition 3.6 once more with

$$\xi_Z = \frac{\xi_A + \xi_B}{\alpha_1\alpha_2}, \quad \xi_{Z(A \otimes B)} = 2(\xi_A + \xi_B),$$

we obtain an $(\alpha_1\alpha_2, a_1 + a_2 + 4\log(n) + 6, 3(\xi_A + \xi_B))$-block-encoding of $Z(A \otimes B)Z$.

Just as was the case with Z, we can use the description of G to construct the sparse-access oracles O_r and O_c as defined in Lemma 3.2 (which again, act on two $(2\log n + 1)$ qubit registers), as well as an oracle O_G using $\tilde{\mathcal{O}}_n(1)$ gates, that returns the entries of G in a binary description:

$$O_G : |i\rangle|j\rangle|0\rangle^{\otimes p} \mapsto |i\rangle|j\rangle|g_{ij}\rangle, \quad \forall i,j \in [2^{2\log n}] - 1,$$

where g_{ij} is a p-bit binary description of G_{ij} (the ij-matrix element of G). Noting that G is 1-row sparse and 1-column sparse (and hence, so its transpose); applying Lemma 3.2 twice more allows us to construct a $(1, 2\log(n) + 3, \xi_G)$-block-encoding U_G of G, as well as a $(1, 2\log(n) + 3, \xi_G^\top)$-block-encoding $U_{G^\top}$ of the transpose $G^\top$. We can then use U_G and our $(\alpha_1\alpha_2, a_1 + a_2 + 4\log(n) + 6, 3(\xi_A + \xi_B))$-block-encoding $U_{Z(A\otimes B)Z}$ of $Z(A\otimes B)Z$ to construct an $(\alpha_1\alpha_2, a_1 + a_2 + 6\log(n) + 9, 4(\xi_A + \xi_B))$-block-encoding of $G(Z(A\otimes B)Z)$ by applying Proposition 3.6 with

$$\xi_G = \frac{\xi_A + \xi_B}{\alpha_1\alpha_2}, \quad \xi_{Z(A\otimes B)Z} = 3(\xi_A + \xi_B).$$

Applying Proposition 3.6 a final time, with

$$\xi_{G^\top} = \frac{\xi_A + \xi_B}{\alpha_1\alpha_2}, \quad \xi_{G(Z(A\otimes B)Z)} = 4(\xi_A + \xi_B),$$

produces an $(\alpha_1\alpha_2, a_1 + a_2 + 8\log(n) + 12, 5(\xi_A + \xi_B))$-block-encoding $U_{\mathcal{M}}$ of $\mathcal{M} = G(Z(A\otimes B)Z)G^\top$.

The stated complexity result follows upon noting that the steps required to construct the unitary

$$U_{\mathcal{M}} = U_G U_Z U_{A\otimes B} U_Z U_{G^\top}$$

requires one application of $U_{A\otimes B}$ and one application of each of the other matrices. In turn, this amounts to 1 application of U_A and U_B each, plus the $\tilde{\mathcal{O}}_n(1)$ gate cost of the remaining matrices U_G, U_Z and $U_{G^\top}$, and the proof is complete. $\qquad\square$

We remark that a similar result to Proposition 3.11 was independently derived and discussed in the recent paper [CCW22]. We point out that the construction provided in [CCW22] is less explicit than what we provide here; they simply note that the Hadamard product can be obtained as a reduction of the tensor product, and they do not consider the scaling of the error of the block-encoding in their analysis.

3.3.5 Implementing powers of matrices

The next two propositions indicate that if one has a block-encoded matrix, one can implement a block-encoding of powers of that matrix – most notably, we can implement the inverse matrix.

Proposition 3.12 (Lemma 9 in [CGJ19]). *(Implementing negative powers of Hermitian matrices) Let $r \in (0, \infty)$, $\kappa \geq 2$ and H a Hermitian matrix such that $I/\kappa \preceq H \preceq I$. Suppose that*

$$\delta = o\left(\frac{\epsilon}{\kappa^{1+r}(1+r)\log^3 \frac{\kappa^{1+r}}{\epsilon}}\right)$$

and U is an (α, a, δ)-block-encoding of H, that can be implemented in time T_U. Then, for any ϵ, we can implement a unitary $\tilde{U}$ which is a $(2\kappa^r, a + \mathcal{O}(\log(\kappa^{1+r}\log 1/\epsilon), \epsilon)$-block-encoding of H^{-r} in time

$$\mathcal{O}\left(\alpha\kappa(a + T_U)(1+r)\log^2\left(\frac{\kappa^{1+r}}{\epsilon}\right)\right).$$

Proposition 3.13 (Lemma 10 in [CGJ19]). *(Implementing positive powers of Hermitian matrices) Let $r \in (0,1]$, $\kappa \geq 2$ and H a Hermitian matrix such that $I/\kappa \preceq H \preceq I$. Suppose that*

$$\delta = o\left(\frac{\epsilon}{\kappa \log^3 \frac{\kappa}{\epsilon}}\right)$$

and U is an (α, a, δ)-block-encoding of H, that can be implemented in time T_U. Then, for any ϵ, we can implement a unitary $\tilde{U}$ which is a $(2, a + \mathcal{O}(\log\log(1/\epsilon)), \epsilon)$-block-encoding of H^r in time

$$\mathcal{O}\left(\alpha\kappa(a + T_U)\log^2\left(\frac{\kappa}{\epsilon}\right)\right).$$

These results are based on ideas from van Apeldoorn et al. [vAGGdW20], which demonstrate how one can implement smooth functions of a Hamiltonian H based on Fourier series decompositions. We omit a discussion on those techniques here, as we discuss an improved framework for implementing sooth functions of Hermitian matrices next.

3.4   Quantum Singular Value Transformation

In the preceding section, we examined the construction for a number of basic linear algebraic operations. While it is true that any smooth function can be implemented through an aptly chosen combination of these basic operations, a new paradigm from Gilyén et al. [GSLW19] termed *quantum singular value transformation (QSVT)*, offers a more elegant approach. Not only does QSVT provide a unifying framework for many quantum algorithms, it also allows one to perform singular value estimation and implement functions of Hermitian matrices. These developments led to quadratic speedups for Grover search, amplitude estimation and quantum walks, while also yielding exponential speedups for Hamiltonian simulation, all of which frequently arise in the design of quantum algorithms.

Suppose that our matrix A has singular values $\sigma_1, \ldots, \sigma_n$. Without loss of generality, we assume access to a degree-d odd polynomial map $P : [-1,1] \mapsto [-1,1]$, and that we have access to a description of A in its singular value decomposition form, i.e.:

$$U = \begin{pmatrix} A & \cdot \\ \cdot & \cdot \end{pmatrix} = \begin{pmatrix} \sum_i \sigma_i |w_i\rangle\langle v_i| & \cdot \\ & \cdot \end{pmatrix},$$

where $|w_i\rangle$ and $|v_i\rangle$ are the singular vectors. Then, using P, we can implement the unitary

$$U_\Phi = \begin{pmatrix} \sum_i P(\sigma_i)|w_i\rangle\langle v_i| & \cdot \\ & \cdot \end{pmatrix},$$

which stores a matrix in which the singular values have been transformed, but the singular vectors are left unchanged. Moreover, the authors in [GSLW19] point out that for Hermitian matrices the singular value transformation and eigenvalue transformation coincide.

Using this foundation, one can improve upon the results in Propositions 3.12 and 3.13.

Proposition 3.14 (Corollary 3.4.13 in [Gil19]). *(Polynomial approximations of negative power functions) Let $\delta, \epsilon \in (0, 1/2]$, $c > 0$ and let $f(x) = \frac{\delta^c}{2} x^{-c}$, then there exist even/odd polynomials $P, P' \in \mathbb{R}[x]$ such that $\|P - f\|_{[\delta,1]} \leq \epsilon$, $\|P\|_{[-1,1]} \leq 1$, and similarly $\|P' - f\|_{[\delta,1]} \leq \epsilon$, $\|P'\|_{[-1,1]} \leq 1$, moreover the degrees of the polynomials are $\mathcal{O}\left(\frac{\max[1,c]}{\delta} \log \left(\frac{1}{\epsilon} \right) \right)$.*

Proposition 3.15 (Corollary 3.4.13 in [Gil19]). *(Polynomial approximations of negative power functions) Let $\delta, \epsilon \in (0, 1/2]$, $c > 0$ and let $f(x) = \frac{\delta^c}{2} x^{-c}$, then there exist even/odd polynomials $P, P' \in \mathbb{R}[x]$ such that $\|P - f\|_{[\delta,1]} \leq \epsilon$, $\|P\|_{[-1,1]} \leq 1$, and similarly $\|P' - f\|_{[\delta,1]} \leq \epsilon$, $\|P'\|_{[-1,1]} \leq 1$, moreover the degrees of the polynomials are $\mathcal{O}\left(\frac{\max[1,c]}{\delta} \log \left(\frac{1}{\epsilon} \right) \right)$.*

3.4.1 Gibbs Samplers and Trace Estimators

Using QSVT, we can also construct a Gibbs sampler and a trace estimator.

Definition 3.3 (Definition 4.11 in [vA20]). *A θ-precise Gibbs-sampler for the input matrix H, is a unitary that takes as input a data structure storing a Hamiltonian H and creates as output a purification of a θ-approximation (in trace distance) of the Gibbs state*

$$\rho = \frac{\exp(-H)}{\mathrm{tr}(\exp(-H))}.$$

We will use these approximate Gibbs states in order to check the diagonal entries of our solutions, as well as compute the trace inner products of matrices (or, expectation values), i.e., quantities of the form $\mathrm{tr}(A\rho)$.

Definition 3.4 (Definition 4.12 in [vA20]). *A θ-precise trace estimator is a unitary that as input takes a state ρ and a matrix A. It outputs a sample from a random variable $X \in \mathbb{R}$ such that X is an estimator for $\mathrm{tr}(A\rho)$ that is at most $\theta/4$ biased.*

These implementations require polynomial approximations of the exponential function, which can be obtained using quantum singular value transformation techniques introduced in [Gil19, GSLW19].

Lemma 3.16 (Lemma 4.14 in [vA20]). *Let $\xi \in (0, 1/6]$ and $\beta \geq 1$. There exists a polynomial $P(x)$ such that*

- *For all $x \in [-1, 0]$, we have $\left| P(x) - \exp(2\beta x)/4 \right| \leq \xi$.*

- *For all $x \in [-1, 1]$, we have $\left| P(x) \right| \leq 1/2$.*

- *$\deg(P) = \widetilde{\mathcal{O}}_{\frac{1}{\xi}}(\beta)$.*

Proof. We refer the reader to the proof of Corollary 64 in [GSLW19]. $\qquad\square$

Lemma 3.17 (Lemma 4.15 in [vA20]). *Let $\theta \in (0, 1/3]$, $\beta > 1$, and let d be the degree of the polynomial from Lemma 3.16 when we let $\xi = \frac{\theta}{128n}$. Let U be a $(\beta, a, \frac{\theta^2 \beta}{1024^2 d^2 n^2})$-block encoding* of a Hermitian operator $H \in \mathbb{R}^{n \times n}$, i.e., a $(\beta, a, \tilde{\mathcal{O}}(\theta/\beta n^2))$-block encoding. Then, we can create a purification of a state $\tilde{\rho}$ such that

$$\left\| \tilde{\rho} - \frac{\exp(H)}{\operatorname{tr}\left(\exp(H)\right)} \right\|_{\mathrm{tr}} \leq \theta$$

using $\tilde{\mathcal{O}}_{\frac{1}{\theta}}(\sqrt{n}\beta)$ applications of U and $\tilde{\mathcal{O}}_{\frac{1}{\theta}}(\sqrt{n}\beta a)$ elementary operations.

Provided access to a unitary that prepares a purification of a density operator, we can also construct a block-encoding of it. This is formalized in the following lemma found in [Gil19], which was based on ideas of [LC19, Corollary 9].

Lemma 3.18 (Lemma 6.4.4). *(Block-encoding of a (subnormalized) density operator) Let G be a $(w+a)$ unitary which on the input state $|0\rangle^{w}|0\rangle^{a}$ prepares a purification $|\varrho\rangle$ of the subnormalized w-qubit density operator ϱ. Then we can implement a $(1, w + a, 0)$-block-encoding of ϱ with a single use of G and its inverse and with $w + 1$ two-qubit gates.*

We are now in a position to define a trace estimator using the quantum operator input model.

Lemma 3.19 (Lemma 4.18 in [vA20]). *Let ρ be an n-dimensional quantum state and U an $(\alpha, a, \theta/2)$-block encoding of a matrix $A \in \mathbb{R}^{n \times n}$ with $\|A\| \leq 1$. A trace estimator for $\operatorname{tr}(A\rho)$ with bias at most θ and $\sigma = \mathcal{O}(1)$ can be implemented using $\tilde{\mathcal{O}}(\alpha)$ uses of U and $U^{\dagger}$ and $\tilde{\mathcal{O}}_{\frac{1}{\theta}}(\alpha)$ elementary operations.*

Next, we see how these results can be utilized to prepare a quantum state that encodes the solution to a linear system of equations.

3.5 Solving linear systems of equations

As we saw earlier, if one has a block-encoded matrix, then one can also implement a block-encoding of negative powers of that matrix, (see, Propositions 3.12 and 3.14). This provides an framework for solving linear systems of equations.

Letting U be a (α, a, δ)-block encoding of A that can be implemented in time T_U. Then, we can implement a block-encoding V of the inverse of A:

$$V = \begin{pmatrix} \frac{A^{-1}}{2\kappa} & \cdot \\ \cdot & \cdot \end{pmatrix}$$

Hence, given A as a block-encoding and $|b\rangle$, we can solve the linear system

$$Ax = b$$

by outputting the state

$$\frac{A^{-1}|b\rangle}{\|A^{-1}|b\rangle\|}$$

In particular, we apply a technique known as amplitude amplification to the state:

$$V|u\rangle|0\rangle^{\otimes a} = \frac{1}{2\kappa}A^{-1}|u\rangle|0\rangle^{\otimes a} + |\cdot\rangle.$$

We are now in a position to formalize how block-encodings solve quantum linear systems problems [CGJ19].

Theorem 3.20 (Theorem 30 in [CGJ19]). *(Solution of linear system) Let $r \in (0,\infty)$, $\kappa \geq 2$ and H a Hermitian matrix such that its nonzero eigenvalues lie in $[-1,-1/\kappa] \cup [1/\kappa, 1]$. Suppose that*

$$\delta = o\left(\frac{\xi}{\kappa^2 \log^3 \frac{\kappa^2}{\xi}}\right)$$

and U is an (α, a, δ)-block-encoding of H, that can be implemented in time T_U. Suppose further that we can prepare a state $|v\rangle$ that is in the image of H in time T_v. Then, for any ξ, we can output a state that is ξ-close to $H^{-1}|v\rangle/\|H^{-1}v\|$ in time

$$\mathcal{O}\left(\kappa\left(\alpha(a + T_U)\log^2\left(\frac{\kappa}{\xi}\right) + T_v\right)\log \kappa\right).$$

QLSAs can prepare a state proportional to our solution in time that is both polylogarithmic in the dimension of our linear system and the inverse of the precision $1/\delta$. We can also output an estimate of the norm of the solution by increasing the running time of the previous result by a factor of $1/\delta$.

Proposition 3.21 (Corollary 32 in [CGJ19]). *(Norm estimation) Let $p \in (0,\infty)$, $\kappa \geq 2$ and H a Hermitian matrix such that its nonzero eigenvalues lie in $[-1,-1/\kappa] \cup [1/\kappa, 1]$. Suppose that*

$$\delta = o\left(\frac{\xi}{\kappa^2 \log^3 \frac{\kappa^2}{\xi}}\right)$$

and U is an (α, a, δ)-block-encoding of H, that can be implemented in time T_U. Suppose further that we can prepare a state $|v\rangle$ that is in the image of H in time T_v. Then we can output $\tilde{e}$ such that

$$(1 - \xi)\|H^{-1}|v\rangle\| \leq \tilde{e} \leq (1 + \xi)\|H^{-1}|v\rangle\|$$

in time

$$\mathcal{O}\left(\frac{\kappa}{\xi}\left(\alpha(a + T_U)\log^2\left(\frac{\kappa}{\xi}\right) + T_v\right)\log^3 \kappa \log\frac{\log \kappa}{\delta}\right).$$

As we have stressed, QLSAs prepare a *quantum state* that encodes the solution to a linear system. However, a naïve application of QLSA in combination with a quantum tomography algorithm to learn the result results leads to linear dependence on the dimension and the condition number, as well as nonpolynomial dependence on the inverse precision to which we solve our system (whereas inexact classical linear systems algorithms depend polylogarithmically on this quantity). This is further compounded by the fact that we need to scale down our data due to the normalization of quantum states, which accordingly appears in the running time as part of the error scaling. Given the fact that, at the present time of writing, we simply have more use for classical information than information encoded in a quantum state, it is of interest to mitigate this scaling in the error. This is what we discuss in the next chapter.

Chapter 4

Iterative Refinement for Linear Systems

This chapter is based on the paper [MAF+22].

Recently, a considerable amount of attention has been devoted to quantum linear algebra; with a particular interest in a class of methods that use Hamiltonian simulation subroutines to prepare a quantum state $|x\rangle$ that is proportional to the solution of the linear system

$$Ax = b,$$

given a matrix $A \in \mathbb{R}^{d \times d}$ and a vector $b \in \mathbb{R}^d$. In particular, quantum computers have been shown to be able to prepare $|x\rangle$ in time that is polylogarithmic in the dimension. However, learning a classical description of the solution requires the use of quantum state tomography, naturally introduces polynomial overhead in the inverse precision. In this chapter, we study how using an iterative refinement algorithm allows one to recover polylogarithmic scaling in the error even when a classical description of the solution is desired.

4.1 Prior work

Research into the subfield of solving linear systems of equations on a quantum computer began with the work of Harrow, Hassidim, and Lloyd [HHL09], who proposed what has come to be known as the HHL algorithm for solving the *quantum linear systems problem* (QLSP). In this seminal work, it was shown that a quantum computer could be used to solve a QLSP with a worst-case complexity of

$$\mathcal{O}\left(\frac{s^2 \kappa_A^2}{\epsilon} \cdot \mathrm{polylog}(d)\right),$$

where ϵ is the accuracy to which the solution is obtained. In the standard setting, it is assumed A has at most s nonzero entries per row, satisfies $\|A\| \leq 1$ and that the nonnegative eigenvalues of A are assumed to lie in the interval $[-1, -1/\kappa_A] \cup [1/\kappa_A, 1]$ such that $\|A^{-1}\| \leq \kappa_A$. Moreover, it is generally assumed we have oracles which provide access to the entries of A, and the ability to prepare a quantum state $|b\rangle$ that is proportional to the right hand side vector b.

Research into this subfield began with the work of Harrow, Hassidim, and Lloyd [HHL09], who proposed what has come to be known as the HHL algorithm for solving the *quantum linear systems problem* (QLSP). Although the HHL algorithm exhibited quadratic dependence on κ_A, its polylogarithmic dependence to the dimension of the problem opened the doors for a potential exponential quantum speedup for solving linear systems of equations.

Given the fundamental role that linear systems play in science and mathematics, the paper [HHL09] led to a series of works focused on improving the worst-case running time of quantum linear systems algorithms (QLSAs). The first enhancements to the complexity of the HHL algorithm were made in [Amb12, CKS17, SSO19], in which the dependence on κ_A was improved from quadratic to almost linear. QLSAs based on variational approaches have also been proposed [see, e.g., AL19, BPLC$^+$19, HBR19, XSE$^+$21].

The HHL algorithm is one of many applications that are built on the idea of using Hamiltonian simulation to simulate the unitary e^{iH} on an input state $|\psi\rangle$ for some time t and a given Hermitian matrix H. To achieve the stated complexity result, the HHL algorithm and its variants utilize the *sparse-access input model*. A different model, now commonly referred to as the *quantum operator input model*, is proposed in [LC19], which is based on the idea of *block-encoded* matrices, which we studied in detail in Chapter 3. QLSAs based on the block-encoding framework were presented in [CKS17, WZP18]. The running time of this class of QLSAs based was subsequently improved by [CGJ19], who develop a technique they call variable-time amplitude *estimation* to obtain a QLSA with worst case complexity

$$\mathcal{O}\left(\kappa_A \alpha \cdot \mathrm{polylog}\left(d, \kappa_A, \frac{1}{\epsilon}\right)\right),$$

where $\alpha = \min\{\|A\|_F, \sqrt{d}\}$ if A is stored in QRAM, with $\alpha = s$ otherwise. It is additionally shown in [CGJ19] how one can efficiently prepare block encodings of a matrix using various input models.

QLSAs prepare a quantum state proportional to the solution to the linear system in time that is polylogarithmic in the dimension, however, our interest usually lies in classically learning the state we obtain. Classically learning the state encoding our solution has historically introduced polynomial overhead due to the need for costly quantum state tomography. Even if we have access to QRAM and use state-of-the art techniques for quantum state tomography, (classically) solving a linear system of equations to precision ϵ under this framework would require time

$$\mathcal{O}\left(\left(d\alpha\kappa_A^2\frac{1}{\epsilon} + d^2\kappa_A\frac{1}{\epsilon}\right) \cdot \text{polylog}\left(d, \kappa_A, \frac{1}{\epsilon}\right)\right).$$

Not only is the above cost asymptotically worse than inexact classical approaches for solving linear systems of equations, but the dependence on the precision implies a non-polynomial overall running time.

Iterative Refinement (IR) is a classical technique for computing extended-precision (or, exact) solutions to linear optimization problems (LOPs) [GS20, GSW12, GSW16], as well as mixed integer optimization problems [ACDE07, CKSW11]. Relevant to the work in this manuscript is the iterative refinement method for solving linear system of equations found in [GVL13], and we summarize the core idea as follows.

The algorithm starts from an initial solution $x^{(0)} \in \mathbb{R}^d$, and subsequently computes a refined solution $x^{(k+1)} \leftarrow x^{(k)} + u^{(k)}$ in iterations $k = 0, 1, 2, \ldots$, where $u^{(k)}$ acts as a correction of the error $r^{(k)} = b - Ax^{(k)}$ and is the solution to the *refining system* $Au^{(k)} = r^{(k)}$. These operations can all be carried out using the same level of accuracy or *fixed precision*. There is also a *mixed precision* approach, in which the residuals $r^{(k)}$ are computed with a higher level of accuracy than that used to compute $u^{(k)}$ [for more details, see, e.g., GVL13, Wil94], but in this chapter we adopt the fixed approach.

4.1.1 Our results

In this chapter, we introduce an Iterative Refinement scheme that allows one to obtain a classical solution to the linear systems problem using a quantum computer. We demonstrate that, irrespective of the chosen input model, the number of refining steps is polylogarithmic in the inverse precision to which we seek to classically solve the linear system at hand.

When the problem data is stored in a quantum-read/classical-write RAM (QRAM), the algorithm we present obtains a ϵ-precise classical solution to a linear system $Ax = b$ with a worst case overall running time of

$$\mathcal{O}\left(d\alpha\kappa_A^2 \cdot \text{polylog}\left(d, \kappa_A, \frac{\|b\|}{\epsilon}\right)\right),$$

accesses to the QRAM and $\mathcal{O}\left(ds \cdot \text{polylog}\left(d, \kappa_A, \frac{\|b\|}{\epsilon}\right)\right)$ classical arithmetic operations, where

d is the dimension, $\alpha \leq \min\left\{\|A\|_F, \sqrt{d}\right\}$ is the subnormalization for the block-encoding of A, and κ_A is an upper bound on the operator norm of A^{-1}. This implies a linear speedup with respect to dimension over existing classical methods for solving linear systems of equations that are indefinite, asymmetric, or both. Without access QRAM, we prove that the algorithm requires at most

$$\mathcal{O}\left(\left(d\left(\sqrt{s}\kappa_A^2\right)^{1+o(1)} + d^2\kappa_A\right)\cdot\mathrm{polylog}\left(d,\kappa_A,\frac{\|b\|}{\epsilon}\right)\right)$$

queries to the oracles describing A, $\mathcal{O}\left(\left(d^2\left(\sqrt{s}\kappa_A^2\right)^{1+o(1)} + d^3\kappa_A\right)\cdot\mathrm{polylog}\left(d,\kappa_A,\frac{\|b\|}{\epsilon}\right)\right)$ additional gates and $\mathcal{O}\left(ds\cdot\mathrm{polylog}\left(d,\kappa_A,\frac{\|b\|}{\epsilon}\right)\right)$ classical arithmetic operations.

4.2   A quantum oracle for the LSP

To motivate the idea at hand, consider the direct approach of classically solving linear systems using quantum computers. First, the problem needs to be normalized such that QLSA can be used to prepare a state proportional to the system. From here, one can apply a quantum state tomography algorithm to classically estimate the result. Letting θ be an upper bound on the size of the solution to the linear system of interest, it is necessary to run the tomography step to precision $\xi = \epsilon/\theta$ if our aim to classically reconstruct the solution to precision ϵ in ℓ_2-norm precision. This is because ϵ is stated in its relative form; as the solution $|x\rangle$ obtained via QLSA is normalized.

Recall that it is assumed $\|A\| \leq 1$ and the nonzero eigenvalues of A lie in $[-1, -1/\kappa_A] \cup [1/\kappa_A, 1]$ such that $\|A^{-1}\| \leq \kappa_A$. Hence, κ_A is an upper bound on the condition number of A whenever $\|A\| = 1$. Under these assumptions, any solution to $Ax = b$ satisfies

$$\|x\| = \left\|A^{-1}b\right\| \leq \left\|A^{-1}\right\|\|b\| \leq \kappa_A\|b\|.$$

Hence, normalizing the right hand side to

$$\tilde{b} := \frac{b}{\|b\|},$$

any solution of the normalized linear system $A\tilde{x} = \tilde{b}$ satisfies $\|\tilde{x}\| \leq \kappa_A$.

Solving $A\tilde{x} = \tilde{b}$ up to precision $\frac{\xi}{\|b\|}$ suffices to solve $Ax = b$ to precision ξ. The main result of this paper shows that within the context of the scheme we present later, performing the QLSA and tomography steps in *fixed precision* (e.g., $\xi = 10^{-2}$) suffices, leading to an exponential speedup in the overall running time with respect to the inverse precision and the normalization factor $\|b\|$.

We begin with an informal statement of the cost of O_{LS} in terms of its subroutines.

Proposition 4.1. *Let T_{LS} represent the time to prepare and solve the linear system using a quantum computer, and denote by $T_{TO}(T_{LS}, \xi)$ the time to perform quantum state tomography*

Algorithm 1 A quantum oracle for classical linear systems

Require: Input oracle access to matrix $A \in \mathbb{R}^{d \times d}$ with $\left\| A^{-1} \right\| \leq \kappa_A$ and vector $\tilde{b} \in \mathbb{R}^d$ with $\|\tilde{b}\| = 1$, accuracy $\xi \in (0,1)$

Implement an $\left(\alpha, a, o\left(\frac{\xi}{\text{poly}(\kappa_A, \log(1/\xi))} \right) \right)$-block-encoding U_A of A

Construct amplitude encoding $|\tilde{b}\rangle$ of the right hand side vector

Prepare a state that is $\frac{\xi}{2}$-close to the solution of $U_A |x\rangle |0\rangle^{\otimes a} = |\tilde{b}\rangle |0\rangle^{\otimes a}$ using a QLSA

Obtain a classical estimate $\tilde{x}$ of $|x_*\rangle$ using QTA with precision $\frac{\xi}{2\kappa_A}$

Output: An ξ-precise classical estimate $\tilde{x}$ of $|x_*\rangle$.

---

to precision ξ. A call to the quantum linear systems oracle O_{LS} has a quantum gate complexity of

$$\mathcal{O}\left(T_{TO}(T_{LS}, \xi) \right).$$

As we saw in our discussion on quantum state tomography, access to QRAM will allow us to perform our extraction in time that is linear in both the dimension and inverse precision. Moreover, access to QRAM data structures will allow us to efficiently prepare quantum states encoding the right hand side, reducing the overhead associated with preparing our linear system on a quantum computer.

Corollary 4.1 (Complexity of O_{LS} with access to QRAM). *Let $A \in \mathbb{R}^{d \times d}$ be a matrix with $\left\| A^{-1} \right\| \leq \kappa_A$ and $\tilde{b} \in \mathbb{R}^d$ be a vector such that $\|\tilde{b}\| = 1$. Suppose that A and $\tilde{b}$ are stored in a QRAM data structure. Fix $\xi \in (0,1)$. Then, the quantum linear systems oracle O_{LS} outputs an ξ-precise (in the ℓ_2-norm) classical estimate of the solution to $A\tilde{x} = \tilde{b}$ using at most*

$$\mathcal{O}\left(d\alpha \kappa_A^2 \frac{1}{\xi} \cdot \text{polylog}\left(d, \kappa_A, \frac{1}{\xi} \right) \right),$$

accesses to the QRAM, where $\alpha \leq \min\left\{ \|A\|_F, \sqrt{d} \right\}$.

Proof. We begin by constructing a $\left(1, \log(d) + 2, o\left(\frac{\xi}{\text{poly}(\kappa_A, \log(1/\xi))} \right) \right)$-block-encoding U_A of A. Since A is stored in QRAM, Proposition 3.3*(ii)* asserts that this construction requires at most $T_A = \tilde{\mathcal{O}}_{d, \kappa_A, \frac{1}{\xi}}(1)$ accesses to the QRAM. Likewise, Theorem 2.3 asserts that when $\tilde{b}$ is stored in a QRAM data structure, a quantum state $|\tilde{b}\rangle$ encoding $\tilde{b}$ can be prepared using $T_{\tilde{b}} = \tilde{\mathcal{O}}_{\frac{d}{\xi}}(1)$ queries to the QRAM.

From here, applying Proposition 3.3*(ii)* together with Theorem 3.20 implies that we can prepare a state that is $\frac{\xi}{2}$-close to a state $|x\rangle$ which solves $U_A |x\rangle = |\tilde{b}\rangle$ using at most

$$T_{LS}^{\text{QRAM}} = \mathcal{O}\left(\kappa_A \left(\alpha \left(\log(d) + T_A \right) \log^2\left(\frac{\kappa_A}{\xi} \right) + T_{\tilde{b}} \right) \log(\kappa_A) \right) = \tilde{\mathcal{O}}_{d, \kappa_A, \frac{1}{\xi}} (\alpha \kappa_A)$$

accesses to the QRAM, as $T_A = \tilde{\mathcal{O}}_{d, \frac{\kappa_A}{\xi}}(1)$ and $T_{\tilde{b}} = \tilde{\mathcal{O}}_{\frac{d}{\xi}}(1)$. Applying Theorem 2.5 with $T_U = T_{LS}^{\text{QRAM}}$ and noting that the precision of the tomography step is chosen to be $\frac{\xi}{2\kappa_A}$, it follows that when the problem data is stored in QRAM, the oracle O_{LS} can output an ξ-precise (in the ℓ_2-norm) classical solution to $Ax = \tilde{b}$ using at most

$$\mathcal{O}\left(T_{TO}\left(T_{LS}^{\text{QRAM}}, \frac{\xi}{\kappa_A}\right)\right) = \tilde{\mathcal{O}}_{d,\frac{1}{\xi}}\left(T_{LS}^{\text{QRAM}} \frac{d}{\xi/\kappa_A}\right) = \tilde{\mathcal{O}}_{d,\kappa_A,\frac{1}{\xi}}\left(d\alpha\kappa_A^2 \frac{1}{\xi}\right)$$

accesses to the QRAM. The proof is complete. $\qquad\qquad\square$

Within the scheme we present later, it can be preferable to take on additional dependence on the inverse precision ξ^{-1} if it allows us to mitigate the dependence on other parameters such as d and s; using an IR framework will reduce this added cost in ξ^{-1} to constant overhead. Along this line, our results in the sparse-access model will rely on the following result from [Low19], which will allow us to save a $\sqrt{s}$ factor in the overall running time that we would otherwise incur using the QLSA from [CGJ19] (which is exponentially faster in the inverse precision).

Theorem 4.2 (Corollary 5 in [Low19]). *(Query complexity of solving sparse linear systems of equations). Let $M \in \mathbb{C}^{N \times N}$ be an s-sparse matrix with $\|M^{-1}\| \leq \kappa_M$. Let $|z\rangle$ be a state prepared by a unitary oracle $O_z|0\rangle = |z\rangle$. Then, the query complexity to all oracles for preparing a state $|x\rangle$ that is ξ-close to $M^{-1}|z\rangle/\|M^{-1}z\|$ using $\mathcal{O}\left(\kappa_M\sqrt{s}\left(\frac{\kappa_M s}{\xi}\right)^{o(1)}\right)$ queries to O_z and the oracles describing M.*

Corollary 4.2 (Complexity of O_{LS} with sparse-access problem data). *Let $A \in \mathbb{R}^{d \times d}$ be an s-sparse matrix with $\|A^{-1}\| \leq \kappa_A$, and $\tilde{b} \in \mathbb{R}^d$ be a vector with norm at most one. Assume sparse oracle access to A, and access to an oracle $O_{\tilde{b}}|0\rangle = |\tilde{b}\rangle$ that prepares a quantum state encoding the right hand side vector $\tilde{b}$. Fix $\xi \in (0,1)$. Then, the quantum linear systems oracle O_{LS} outputs an ξ-precise (in the ℓ_2-norm) classical estimate of the solution to $A\tilde{x} = \tilde{b}$ using at most*

$$\mathcal{O}\left(\left(d\left(\frac{\kappa_A^2\sqrt{s}}{\xi}\right)^{1+o(1)} + \frac{d^2\kappa_A}{\xi}\right) \cdot \text{polylog}\left(d, \kappa_A, \frac{1}{\xi}\right)\right)$$

queries to $O_{\tilde{b}}$ and the oracles describing A.

Proof. With sparse-oracle access to A and access to an oracle $O_{\tilde{b}}|0\rangle = |\tilde{b}\rangle$ that prepares a quantum state encoding the right hand side vector $\tilde{b}$, Theorem 4.2 asserts that we can prepare a quantum state $|x\rangle$ that is $\frac{\xi}{2\kappa_A}$-close to a quantum state encoding the exact solution to our linear system using

$$T_{LS}^{\text{sparse}} = \mathcal{O}\left(\kappa_A\sqrt{s}\left(\frac{\kappa_A s}{\xi}\right)^{o(1)}\right)$$

queries to $O_{\tilde{b}}$ and the oracles describing A.

Applying Theorem 2.4 with $T_U = T_{LS}^{\text{sparse}}$, it follows that the oracle O_{LS} requires at most

$$\mathcal{O}\left(\left(T_{LS}^{\text{sparse}}\frac{d}{\xi/\kappa_A} + \frac{d^2}{\xi/\kappa_A}\right) \cdot \text{polylog}\left(d, \frac{1}{\xi/\kappa_A}\right)\right)$$

$$= \mathcal{O}\left(\left(d\left(\frac{\kappa_A^2\sqrt{s}}{\xi}\right)^{1+o(1)} + \frac{d^2\kappa_A}{\xi}\right) \cdot \text{polylog}\left(d, \kappa_A, \frac{1}{\xi}\right)\right)$$

queries to $O_{\tilde{b}}$ and the oracles describing A in order to output an ξ-precise classical solution to $Ax = \tilde{b}$, and the proof is complete. $\qquad\square$

4.3 Iterative Refinement for the LSP using quantum subroutines

Iterative refinement is a classical computing technique that is used to obtain accurate solutions to linear systems of equations. In this section, we summarize the core idea at high level, before providing theoretical results for our algorithm.

4.3.1 The algorithm

Our Iterative Refinement method for the LSP is presented in detail in Algorithm 2. The algorithm takes as input the linear system coefficient matrix A, right hand side vector b, and in every iteration there is a call to the quantum linear systems oracle O_{LS} which we described in detail in Section 4.2. There are two precision parameters: ξ, the fixed precision we use for each of our oracle calls, and ζ, the desired precision of the final solution.

The algorithm commences from an initial point $x^{(0)} \in \mathbb{R}^d$ which need not necessarily satisfy

$$Ax^{(0)} \approx \tilde{b}.$$

In each iteration, we call O_{LS} to solve the refining system

$$Au^{(k)} \approx \eta^{(k)} r^{(k)}, \tag{4.1}$$

where $\eta^{(k)} \geq 1$ is a scaling factor, and

$$r^{(k)} = \tilde{b} - Ax^{(k)}, \tag{4.2}$$

is the error associated with the current solution. A solution $u^{(k)}$ of (4.1) therefore acts as a correction for the error $r^{(k)}$ and we use this quantity to update our solution to the linear system $Ax = \tilde{b}$ as follows:

$$x^{(k+1)} \leftarrow x^{(k)} + \frac{1}{\eta^{(k)}}u^{(k)}.$$

Given that we construct O_{LS} using quantum subroutines, we must properly set bounds on the error tolerances for these steps. The aim of this section is to demonstrate that our Iterative Refinement scheme allows one to simply utilize a fixed level of precision (e.g., $\xi = 10^{-2}$) for every call to O_{LS} in Algorithm 2. These operations can all be carried out using the same level of accuracy or *fixed precision*, which is the approach we take in this work. Once the norm of the error r has been reduced to or below a desired precision, i.e., $\|r^{(k)}\| \leq \zeta$ for some $0 < \zeta \ll \xi$, the algorithm terminates and reports $x^{(k)}$ as the solution.

Algorithm 2 Iterative Refinement for the Linear Systems Problem from [MAF$^+$22]

Input: Coefficient matrix $A \in \mathbb{R}^{d \times d}$ with $\|A^{-1}\| \leq \kappa_A$, right hand side vector $b \in \mathbb{R}^d$, error tolerances $0 < \zeta \ll \xi < 1$.

Output: ζ-precise classical solution vector x to the linear system $Ax = \frac{b}{\|b\|}$.

Normalize right hand side $\tilde{b} \leftarrow \frac{b}{\|b\|}$.

Choose starting point $(x^{(0)}, r^{(0)}, \eta^{(0)}) \leftarrow (0, \tilde{b}, 1)$, $k \leftarrow 0$.

while $\|r^{(k)}\| > \zeta$

    1. $\tilde{u}^{(k)} \leftarrow$ **solve** $(A, \eta^{(k)} r^{(k)})$ using $O_{LS}(\xi)$

    2. $x^{(k+1)} \leftarrow x^{(k)} + \frac{1}{\eta^{(k)}} \tilde{u}^{(k)}$

    3. $r^{(k+1)} \leftarrow \tilde{b} - Ax^{(k+1)}$

    4. $\eta^{(k+1)} \leftarrow \frac{1}{\|r^{(k+1)}\|}$

    5. $k \leftarrow k + 1$

end

4.3.2 Convergence

First, we establish that the solution obtained in Step 1 is a ξ-precise classical solution (in the ℓ_2-norm) to the refining problem at iteration k.

Proposition 4.3. *Let $u^{(k)}$ be the solution to the linear system $Au^{(k)} = \eta^{(k)} r^{(k)}$. Then, for all $k \geq 1$, it follows*

$$\left\| u^{(k)} \right\| \leq \kappa_A.$$

Proof. We begin by establishing the result holds for $k = 0$. Note that we initialize $\eta^{(0)} = 1$ and $r^{(0)} = \tilde{b} = \frac{b}{\|b\|}$. Hence,

$$\left\| u^{(0)} \right\| = \left\| A^{-1} \eta^{(0)} r^{(0)} \right\| \leq \left\| A^{-1} \right\| \left\| \tilde{b} \right\| \leq \kappa_A.$$

Next, note that for all $k \geq 1$ the inequality

$$\left\| \eta^{(k)} r^{(k)} \right\| \leq \eta^{(k)} \left\| r^{(k)} \right\| = 1$$

trivially holds, since we define $\eta^{(k)} = \left\|r^{(k)}\right\|^{-1}$. Thus, repeating a similar argument asserts

$$\left\|u^{(k)}\right\| \leq \kappa_A$$

for $k \geq 1$. The proof is complete. $\qquad\square$

Corollary 4.3. *Let $\tilde{u}^{(k)}$ the classical vector obtained from using a ξ-precise call to O_{LS} with input data $(A, \eta^{(k)}r^{(k)})$. For all $k \geq 0$, we have:*

$$\left\|\eta^{(k)}r^{(k)} - A\tilde{u}^{(k)}\right\| \leq \xi.$$

In other words, $\tilde{u}^{(k)}$ is an ξ-precise classical solution to the refining system $Au^{(k)} = \eta^{(k)}r^{(k)}$.

Proof. The result is a direct consequence of Proposition 4.3 and the definition of the oracle O_{LS}. Since norms of the solutions encountered during the algorithm are at most κ_A, and the tomography steps in O_{LS} are performed to precision $\frac{\xi}{2\kappa_A}$, the desired result follows from a simple triangle inequality. $\qquad\square$

In the next result, we provide a lower bound on the value of the scaling factor at iteration k, $\eta^{(k)}$, and establish that the sequence of solutions generated by Algorithm 2 satisfies the normalized linear system $Ax = \tilde{b}$ with increasing accuracy.

Theorem 4.4. *Let $x^{(k)}$ and $\eta^{(k)}$, $k = 0, 1, \ldots$ be the sequence of solutions and scaling factors produced by Algorithm 2. Then for all $k \geq 0$,*

(a) $\eta^{(k)} \geq \frac{1}{\xi^k}$,

(b) $\left\|\tilde{b} - Ax^{(k)}\right\| \leq \xi^k$.

Proof. To prove *(a)*, we can follow a simple proof by induction. First, observe that $\eta^{(k)} \geq \frac{1}{\xi^k}$ trivially holds at $k = 0$ due to initializing $\eta^{(0)} = 1 = \frac{1}{\xi^0}$. Having shown that the relation holds at $k = 0$, we now assume that $\eta^{(\ell)} = \frac{1}{\xi^\ell}$ is true for $\ell = 1, \ldots, k$ under the induction hypothesis.

In order to complete the induction argument, we demonstrate that the precision of the solution is $\frac{\xi}{\eta^{(k)}}$ at iteration k. Applying Corollary 4.3, the classical estimate $\tilde{u}^{(k)}$ satisfies

$$\|\eta^{(k)}r^{(k)} - A\tilde{u}^{(k)}\| \leq \xi.$$

It therefore follows that the updated solution in each iteration satisfies

$$\|r^{(k+1)}\| = \left\|\tilde{b} - Ax^{(k+1)}\right\| = \left\|\tilde{b} - A\left(x^{(k)} + \frac{1}{\eta^{(k)}}\tilde{u}^{(k)}\right)\right\| = \left\|\tilde{b} - Ax^{(k)} - A\frac{1}{\eta^{(k)}}\tilde{u}^{(k)}\right\|$$

$$= \frac{1}{\eta^{(k)}}\left\|\eta^{(k)}r^{(k)} - A\tilde{u}^{(k)}\right\| \leq \frac{\xi}{\eta^{(k)}}.$$

That is, $x^{(k+1)} = x^{(k)} + \frac{1}{\eta^{(k)}}\tilde{u}^{(k)}$ is an $\frac{\xi}{\eta^{(k)}}$-precise solution of the linear system $Ax = \tilde{b}$ for $k \geq 0$. Noting that $\|r^{(k+1)}\| \leq \frac{\xi}{\eta^{(k)}}$ for all $k \geq 0$, it follows

$$\frac{1}{\|r^{(k+1)}\|} \geq \frac{1}{\frac{\xi}{\eta^{(k)}}} = \frac{\eta^{(k)}}{\xi}.$$

Since $\eta^{(k)} \geq \frac{1}{\xi^k}$ holds under the induction hypothesis, we have

$$\eta^{(k+1)} = \frac{1}{\|r^{(k)}\|} \geq \frac{\eta^{(k)}}{\xi} \geq \frac{\frac{1}{\xi^k}}{\xi} = \frac{1}{\xi^{k+1}}.$$

The result in *(b)* follows trivially from the result in *(a)*, and the fact that the precision of the solution at iteration k is $\frac{\xi}{\eta^{(k)}}$, which together yield

$$\|r^{(k)}\| \leq \frac{\xi}{\eta^{(k-1)}} \leq \frac{\xi}{\frac{1}{\xi^{k-1}}} \leq \xi^k.$$

$\square$

The next result establishes polynomial convergence of Algorithm 2.

Corollary 4.4. *Let $0 < \zeta \ll \xi$, and $\eta^{(0)} = 1$. Then, Algorithm 2 terminates in at most $\mathcal{O}\left(\log\left(\frac{1}{\zeta}\right)\right)$ iterations.*

Proof. The result is a direct consequence of Theorem 4.4*(b)*. $\square$

It is important for us to point out that the iteration bound provided in Corollary 4.4 is independent of the chosen input model, and is valid irrespective of whether or not we have access to QRAM. Moreover, it suffices to use fixed values for each of our oracle calls, e.g., one can simply set $\xi = 10^{-2}$.

4.4 Complexity

In this section, we bound the worst-case running time of Algorithm 2 for both models of computation.

4.4.1 Running time with access to QRAM

The following result bounds the per-iteration cost of Algorithm 2 when the problem data is stored in QRAM.

Theorem 4.5. *Suppose that we have classical access to an s-sparse matrix $A \in \mathbb{R}^{d \times d}$ with $\|A\| \leq 1$ and $\|A^{-1}\| \leq \kappa_A$, and a vector $b \in \mathbb{R}^d$. Define $\alpha \leq \min\left\{\|A\|_F, \sqrt{d}\right\}$. Let $\xi \in (0,1)$ be the fixed precision used for the oracle O_{LS} in every iteration and set $\zeta = \frac{\epsilon}{\|b\|}$, where $\epsilon \in (0,1)$ is the precision to which we seek to solve the system $Ax = b$. Then, in the QRAM input model,*

Algorithm 2 classically outputs a ϵ-precise solution to the linear system $Ax = b$ using at most

$$\mathcal{O}\left(d\alpha\kappa_A^2 \cdot \text{polylog}\left(d, \kappa_A, \frac{\|b\|}{\epsilon}\right)\right),$$

accesses to the QRAM, and $\mathcal{O}\left(ds \cdot \text{polylog}\left(d, \kappa_A, \frac{\|b\|}{\epsilon}\right)\right)$ arithmetic operations.

Proof. Now, observe that we can load the problem data (A, b) and normalize the right hand side to

$$\tilde{b} = \frac{b}{\|b\|}$$

using $\mathcal{O}(ds)$ classical arithmetic operations. Observe that solving the normalized $Ax = \tilde{b}$ to precision $\frac{\epsilon}{\|b\|}$ suffices to solve the linear system $Ax = b$ to precision ϵ. Clearly, the oracle O_{LS} also takes κ_A as input (since this is required to construct the QLSA circuit). We would like to avoid naïvely computing κ_A to maintain our speedups, and an efficient approach for computing κ_A is provided later in the proof.

Classical access to A implies that we can store it in QRAM using $\mathcal{O}(ds)$ classical operations, and similarly ηr can be stored in QRAM using $\mathcal{O}(d)$ classical operations. In each iteration of Algorithm 2, we solve the refining linear system using a call to the oracle O_{LS}, and use a classical estimate of this solution to prepare the linear system for the following iterate. We denote the costs of these steps by $T_{O_{LS}}^{\text{QRAM}}$ and $T_{prepare}$, respectively.

By Corollary 4.1, when the problem data is stored in QRAM, an ξ-precise call to O_{LS} requires

$$T_{O_{LS}}^{\text{QRAM}} = \mathcal{O}\left(d\alpha\kappa_A^2\frac{1}{\xi} \cdot \text{polylog}\left(d, \kappa_A\frac{1}{\xi}\right)\right)$$

accesses to the QRAM in the worst-case. Yet, in the context of Algorithm 2, ξ is a fixed constant (e.g., it suffices to set $\xi = 10^{-2}$), and is neglected in $\mathcal{O}$-notation (as it's impact is merely reduced to constant overhead). Hence, the above expression simplifies to

$$T_{O_{LS}}^{\text{QRAM}} = \mathcal{O}\left(d\alpha\kappa_A^2 \cdot \text{polylog}\left(d, \kappa_A\right)\right).$$

From here, we need to use our classical estimate of the solution to the refining problem to update the solution, the residual, and the scaling factor η. It can be readily seen from the pseudocode describing Algorithm 2 that these classical computation steps involve the calculation of matrix-vector products, as well as operations on d-dimensional vectors, such as addition, subtraction and the calculation of the ℓ_2-norm. As A is s-sparse by assumption, each of the required matrix-vector products can be carried out using $\mathcal{O}(ds)$ arithmetic operations, while the vector operations all require $\mathcal{O}(d)$ arithmetic operations. From here, we update the right hand side ηr in the QRAM using $\mathcal{O}(d)$ classical operations. Clearly, the matrix-vector products are the dominant classical operation, implying that

$$T_{prepare} = \mathcal{O}(ds).$$

Summing $T_{O_{LS}}^{\mathrm{QRAM}}$ and $T_{prepare}$, we arrive at the per-iteration cost of Algorithm 2.

In order to efficiently bound κ_A (i.e., without needing to explicitly compute κ_A) one can use the following scheme in the initial iterate of Algorithm 2. Initialize $\kappa_A = 1$ and then

1. Execute steps 1-3 of Algorithm 2.

2. If $\left\| r^{(1)} \right\| \leq \xi$, **accept** κ_A as sufficiently large, and proceed to the next iterate.

3. Otherwise, update $\kappa_A = 2 \cdot \kappa_A$ and **repeat**.

It is trivial to see that it may require $\log_2(\kappa_A)$ iterates of the above scheme in order to set κ_A sufficiently large, therefore only introduces additional polylogarithmic overhead in κ_A into the overall cost of Algorithm 2 (no extraneous computation is required).

Applying Corollary 4.4, it follows that Algorithm 2 terminates in at most $\mathcal{O}\left(\log\left(\frac{1}{\zeta}\right)\right)$ iterations. For our choice of $\zeta = \frac{\epsilon}{\|b\|}$, it follows that in the QRAM input model, an ϵ-precise solution (in ℓ_2-norm) to the LSP can be computed using at most

$$\mathcal{O}\left(d\alpha\kappa_A^2 \cdot \mathrm{polylog}\left(d, \kappa_A^2\right)\log\left(\frac{1}{\zeta}\right)\right) = \mathcal{O}\left(d\alpha\kappa_A^2 \cdot \mathrm{polylog}\left(d, \kappa_A, \frac{\|b\|}{\epsilon}\right)\right)$$

accesses to the QRAM, and $\mathcal{O}\left(ds \cdot \mathrm{polylog}\left(d, \kappa_A, \frac{\|b\|}{\epsilon}\right)\right)$ classical arithmetic operations. The proof is complete. $\qquad\square$

4.4.2 Running time without access to QRAM

The next result gives the total complexity of Algorithm 2 when analyzed in the sparse-access input model.

Theorem 4.6. *Suppose that we have classical access to an s-sparse matrix $A \in \mathbb{R}^{d \times d}$ with $\|A\| \leq 1$ and $\|A^{-1}\| \leq \kappa_A$, and a vector $b \in \mathbb{R}^d$. Let $\xi \in (0, 1)$ be the fixed precision used for the oracle O_{LS} in every iteration and set $\zeta = \frac{\epsilon}{\|b\|}$, where $\epsilon \in (0, 1)$ is the precision to which we seek to solve the system $Ax = b$. Then, using the sparse-access input model, Algorithm 2 classically outputs an ϵ-precise solution of a linear system $Ax = b$ using at most*

$$\mathcal{O}\left(\left(d\left(\sqrt{s}\kappa_A^2\right)^{1+o(1)} + d^2\kappa_A\right) \cdot \mathrm{polylog}\left(d, \kappa_A, \frac{\|b\|}{\epsilon}\right)\right)$$

queries to the oracles describing A,

$$\mathcal{O}\left(\left(d^2\left(\sqrt{s}\kappa_A^2\right)^{1+o(1)} + d^3\kappa_A\right) \cdot \mathrm{polylog}\left(d, \kappa_A, \frac{\|b\|}{\epsilon}\right)\right)$$

Table 4.1: Complexity to classically solve the LSP to ϵ-precision

Algorithm	Complexity	QRAM	Notes
Factorization methods	$\mathcal{O}(d^3)$	-	
Strassen's method [Str69]	$\mathcal{O}(d^{2.81})$	-	
Fast Matrix Multiplication [LG14]	$\mathcal{O}(d^\omega)$	-	$\omega < 2.372864$
Fast sparse solver [PV21]	$\tilde{\mathcal{O}}_{d,\kappa_A,\frac{1}{\epsilon}}\left(d^{2.331645}s\right)$	-	
Conjugate Gradient	$\tilde{\mathcal{O}}_{d,\kappa_A,\frac{1}{\epsilon}}(ds\sqrt{\kappa_A})$	-	$M \succ 0$
Chebyshev polynomials	$\tilde{\mathcal{O}}_{d,\kappa_A,\frac{1}{\epsilon}}(ds\kappa_A)$	-	
QTA+QLSA [CGJ19]	$\tilde{\mathcal{O}}_{d,\kappa_A,\frac{1}{\epsilon}}\left(d\alpha\kappa_A^2\epsilon^{-1}\right)$	✓	$\alpha \le \min\{\|A\|_F, \sqrt{d}\}$
QTA+QLSA [CGJ19]	$\tilde{\mathcal{O}}_{d,\kappa_A,\frac{1}{\epsilon}}\left(ds\kappa_A^2\epsilon^{-1} + d^2\kappa_A\epsilon^{-1}\right)$	✗	
IR-QTA+QLSA (this work)	$\tilde{\mathcal{O}}_{d,\kappa_A,\frac{1}{\epsilon}}\left(d(\alpha\kappa_A^2 + s)\right)$	✓	$\alpha \le \min\{\|A\|_F, \sqrt{d}\}$
IR-QTA+QLSA (this work)	$\tilde{\mathcal{O}}_{d,\kappa_A,\frac{1}{\epsilon}}\left(d^2\left(\sqrt{s}\kappa_A^2\right)^{1+o(1)} + d^3\kappa_A\right)$	✗	

additional gates, and $\mathcal{O}\left(ds \cdot \mathrm{polylog}\left(d, \kappa_A, \frac{\|b\|}{\epsilon}\right)\right)$ *classical arithmetic operations.*

Proof. Upon substituting the application of Corollary 4.1 with Corollary 4.2 and noting that in the absence of QRAM, the oracle $O_{\tilde{b}}$ which prepares the right hand side vector in each iteration (i.e., the vector of residuals) can be constructed using $\tilde{\mathcal{O}}_d(d)$ gates, the proof directly follows the proof of Theorem 4.5. $\qquad\square$

4.4.3 Comparison to existing linear system solvers

Table 4.1 compares the running time of our algorithm to other classical and quantum algorithms in the literature when applied to classically solving linear systems of equations. It can be readily seen that our approach is exponentially faster than the existing quantum approach regardless of whether or not we have access to QRAM. Our algorithms depend polylogarithmically on the inverse of the precision to which we solve the linear system, whereas combining the state of the art QLSA and QTA incurs cost that is polynomial in these quantities.

While CG exhibits a more favorable dependence on the condition number of the coefficient matrix, interestingly, our algorithm with access to QRAM is able to decouple the dependence on the sparsity and the condition number. Thus, while it is unclear whether our approach would offer speedups for problems involving a positive definite coefficient matrix unless $\kappa_A = \mathcal{O}(\mathrm{polylog}(d))$, our methodology would appear to offer an asymptotic speedup over inexact classical solvers whenever $A \not\succ 0$. We emphasize that any speedup over classical algorithms obtained here is reliant on QRAM; there does not appear to be a regime for which our algorithm in the sparse-access input model outperforms both direct and inexact classical approaches simultaneously.

Part II

Quantum Interior Point Methods

Chapter 5

Quantum Interior Point Methods for Semidefinite Optimization

This chapter is based on the publication [ANTZ21] and the paper [MAFT22].

Interior point methods (IPMs) are algorithms to solve convex optimization problems, including semidefinite optimization problems, in polynomial time. The first polynomial-time algorithm for linear optimization (LO) problems, a special class of convex optimization problems, is the ellipsoid method [Kha80]. Karmarkar's IPM improved on the complexity of the ellipsoid method [Kar84]. The first polynomial-time IPM for SDO problems is attributed to Nesterov and Nemirovskii [NN88, NN95]. Their framework employs efficiently-computable self-concordant barrier functions.

The bottleneck of the classical IPM is the solution of a perturbed set of Karush-Kuhn-Tucker (KKT) optimality conditions known as the Newton linear system, which classically may require $\mathcal{O}(n^{2\omega})$ arithmetic operations in practice when applied to SDO problems involving $m = \mathcal{O}(n^2)$ constraints, where $\omega \in [2, 2.38)$ is the matrix-multiplication exponent. Naturally, researchers have begun to study whether QLSAs could be used to accelerate IPMs, but until the work in [ANTZ21], QIPMs for SDO have not been convergent.

5.1 Introduction

In this chapter, we develop quantum interior point methods (QIPMs) for the solution of semidefinite optimization (SDO) problems. Letting $b \in \mathbb{R}^m$, matrices $A_1, \ldots, A_m, C \in \mathcal{S}^n$, where $\mathcal{S}^n$ is the space of $n \times n$ symmetric matrices, we consider the primal SDO problem given as

$$z_P = \inf_X \left\{ \operatorname{tr}(CX) : \operatorname{tr}(A_i X) = b_i, \ \forall i \in [m], X \succeq 0 \right\}. \tag{5.1}$$

We assume that the matrices $A_1, \ldots, A_m$ are linearly independent. The corresponding dual problem is given by

$$z_D = \sup_{y,S} \left\{ b^\top y : \sum_{i=1}^m y_i A_i + S = C, \ S \succeq 0, y \in \mathbb{R}^m \right\}. \tag{5.2}$$

While duality results are stronger for LO than they are for SDO, under the so called *Interior Point Condition* (IPC) strong duality holds. That is, if there is a primal feasible X with $X \succ 0$, and there exists a dual feasible (y, S) with $S \succ 0$, then optimal solutions exist for both the primal and dual SDO problems, and their optimal objective values are equal, i.e., $z_P = z_D$.

The interest in SDO originates from the fact that SDO provides a framework that allows one to formulate fundamental problems in control [BEGFB94], statistics, information theory [Rai01], machine learning [LCB$^+$04, WS06], finance [dEGJL07, WSV12], and quantum information science [Aar19, Eld03, HNW17, Wat09], to name a few. Additionally, SDO can provide tight approximations for various combinatorial optimization problems [GW95, Lov79], and it can be used to study the properties of convex optimization problems [BEGFB94]. SDO problems are attractive since they exhibit practical efficiency [AA00, Stu99, TTT99] and tractability [AHO98, NN88]. LO problems are a special case of SDO problems in which each of the input matrices are diagonal matrices. Considering the vast number of applications of SDO, it is natural to ask if quantum computers can accelerate their solution.

5.1.1 Prior work

The classical literature on algorithms for solving SDO problems is rich and can be categorized into two classes; algorithms that depend polylogarithimically on the precision to which we solve the problem and the size of the minimally inscribed ellipsoid, and algorithms that depend polynomially on these quantities but exhibit an advantage with respect to n and m.

When $m \le \sqrt{n}$, the Cutting Plane Methods of [JLSW20, LSW15] are the best performing classical algorithms for solving SDOPs, and can do so in time

$$\mathcal{O}\left(m(mn^2 + m^2 + n^\omega) \cdot \operatorname{polylog}\left(m, n, R, \frac{1}{\epsilon} \right) \right),$$

where $\omega \in [2, 2.38]$ is the matrix multiplication exponent, R is an upper bound on the trace of primal optimal solutions, ϵ is the precision parameter, and mn^2 is the input size of the SDO problem. However, in practical situations we typically have that m is roughly between n and n^2, in which case the Cutting Plane Methods given in [JLSW20, LSW15] are outperformed by the IPM for SDO from Jiang et al. [JKL$^+$20]. Their IPM exhibits a worst case running time of

$$
\mathcal{O}\left(\sqrt{n}(mn^2 + m^\omega + n^\omega) \cdot \text{polylog}\left(m, n, \frac{1}{\epsilon}\right)\right),
$$

where the term $m^\omega + n^\omega$ represents the per-iteration cost of inverting the Hessian and dual slack matrices. For IPMs, ϵ-optimality implies that the primal and dual feasible solutions exhibit a *normalized* duality gap bounded by ϵ, i.e.:

$$
\frac{\text{tr}(XS)}{n} \le \epsilon.
$$

When $m = \mathcal{O}(n^2)$, the running time of the IPM in [JKL$^+$20] is shown to be $\widetilde{\mathcal{O}}_{\epsilon}\left(n^{5.246}\right)$, providing an enhancement over the running time bound of $\widetilde{\mathcal{O}}_{\epsilon}\left(n^{6.5}\right)$ found in [Mon98, NT97, NT98].

While quantum SDO solvers could also be categorized in a somewhat similar fashion, it is perhaps more natural to do so according to how they attempt to obtain quantum speedups. In this case we also have two classes; at a high level, all proposed quantum SDO solution methodologies quantize a classical algorithm via either a quantum mechanical interpretation of normalized positive semidefinite matrices, or through the use of quantum linear system algorithms (QLSAs) [CGJ19, CKS17, HHL09]. We now review these works in detail.

The first class is comprised of those that quantize algorithms based on matrix exponentials and Gibbs states. The most prominent example is the Matrix Multiplicative Weights Update (MMWU) Method of Arora and Kale [AHK12], which can solve SDO problems in time

$$
\widetilde{\mathcal{O}}_{n, R, \frac{1}{\epsilon}}\left(nms\left(\frac{Rr}{\epsilon_{\text{abs}}}\right)^4 + ns\left(\frac{Rr}{\epsilon_{\text{abs}}}\right)^7\right).
$$

Here, r is an upper bound on the ℓ_1-norm of dual optimal solutions, and ϵ_{abs} is an additive error to which the optimal objective value is approximated. That is, the objective value attained by the acquired solution is OPT $\in [\varsigma - \mathcal{O}(\epsilon_{\text{abs}}), \varsigma + \mathcal{O}(\epsilon_{\text{abs}})]$, where ς is a bound on the optimal objective value determined using binary search. The MWU algorithm alternates between candidate solutions to the primal and dual SDO problems, and does not involve the solution of linear systems.

A quantum SDO solver based on the Arora-Kale framework was constructed by Brandão and Svore [BS17], and van Apeldoorn et al. [vAGGdW20]. With successive improvements [Gri19, vAG19], the running time of the quantum MWU algorithm for SDOPs has been brought down to:

$$
\mathcal{O}\left(\left(\sqrt{m} + \sqrt{n}\frac{Rr}{\epsilon_{\text{abs}}}\right)s\left(\frac{Rr}{\epsilon_{\text{abs}}}\right)^4 \cdot \text{polylog}\left(m, n, R, \frac{1}{\epsilon_{\text{abs}}}\right)\right),
$$

where s is a sparsity parameter (maximum number of non-zero entries per row). While these methods achieve running times with a more favorable dependence on m and n as compared to classical IPM solvers, their polynomial dependence on the parameters R, r and $1/\epsilon_{\mathrm{abs}}$ is exponentially slower than classical IPMs, which depend logarithmically on these quantities. (If ϵ_{abs} is considered to be part of the input, it is expressed in binary format and a running time of $1/\epsilon_{\mathrm{abs}}$ is not polynomial in the usual sense.)

The second class of quantum SDO solvers is comprised of quantum IPMs. This line of research was initiated by Kerenidis and Prakash [KP20], who were the first to investigate whether quantum linear systems algorithms (QLSAs) could be leveraged to speed up the bottleneck of classical IPMs; solving a set of perturbed KKT optimality conditions known as the Newton linear system. However, the algorithm in [KP20] is not convergent; the analysis draws heavily from the classical exact-feasible IPM literature, which operates under the assumption that the Newton linear system is solved *exactly* at each iteration, and consequently primal and dual feasibility are exactly satisfied by the sequence of iterates generated by the algorithm.

Due to the use of quantum state tomography to extract a classical estimate of the state encoding the solution to the Newton system prepared by a QLSA, this cannot be guaranteed without further safeguards. Additional safeguards must be employed to guarantee that the resulting solution will be symmetric; numerous techniques have been proposed for this purpose. These methods are generally grouped into families of search directions such as those introduced by Kojima, Shindoh, and Hara [KSH97], Monteiro and Zhang [Mon96, Mon97, Mon98, Zha98], Monteiro and Tsuchiya [MT99], and Tseng [Tse98]. For an overview of these (as well as many other) search direction, we refer the reader to a study by Todd [Tod99].

The first convergent quantum IPMs (QIPMs) for SDO were presented in [ANTZ21], in which the authors avoid the shortcomings prevalent in early works on QIPMs, by properly symmetrizing the Newton linear system, and using formulations of the Newton linear system that can handle inexact search directions. The first algorithm they propose quantizes an Inexact-Infeasible IPM. Infeasible IPMs were originally motivated by the need of having an initial point (and all subsequent iterates) in the interior of the semidefinite cone. It is easy to have an initial interior point which is infeasible [PS98, Zha94], hence the use of infeasible IPMs. Further, excessive CPU time and memory requirements may prevent solving large linear systems directly. In such a situation, one can employ an iterative method to determine the solutions to the linear systems to compute search directions, and such solutions are not guaranteed to be exact (e.g., due to early stopping of the iterative method). The resulting algorithm is called inexact-infeasible IPM. The first polynomial time inexact-infeasible IPMs were proposed by Korzak [Kor00] and Mizuno and Jarre [MJ99]. Inexact-infeasible IPMs are natural candidates for quantization. Using this terminology, the "prototypical" classical IPM described in [BTN01] can be called an exact-feasible IPM, however, quantizing II-IPMs leads to an algorithm with worse performance in every parameter when compared to classical feasible IPMs. Therefore, we do not discuss II-QIPMs here, as a feasible framework is strictly preferred.

The little-studied inexact-feasible IPM (IF-IPM) bridges the gap between exact-feasible and inexact-infeasible IPMs. In an inexact-feasible IPM, it is typically assumed that primal and dual feasibility can be satisfied exactly, but we allow for a margin of error in solving a reduced Newton system. This is leveraged by the second QIPM framework given in [ANTZ21], which is a novel Inexact-Feasible scheme that utilizes an orthogonal subspace representation of the search directions. This representation guarantees that primal and dual feasibility are satisfied exactly by all the iterates generated by inexact solutions of the Newton linear system obtained via quantum subroutines. Their algorithm converges to an ϵ-optimal solution to an SDO problem involving $n \times n$ matrices and $m = \mathcal{O}(n^2)$ constraints using at most

$$\mathcal{O}\left(n^{3.5}\frac{\kappa^2}{\epsilon} \cdot \mathrm{polylog}\left(n, \kappa, \frac{1}{\epsilon}\right)\right)$$

QRAM accesses and $\mathcal{O}\left(n^{4.5} \cdot \mathrm{polylog}\left(n, \kappa, \frac{1}{\epsilon}\right)\right)$ arithmetic operations.

Although the IF-QIPM achieves a speedup in n over the best performing classical IPMs, though its dependence on an upper bound κ on the condition number of the intermediate Newton linear systems and the precision ϵ to which the SDO problem is solved suggest no quantum advantage overall: the complexity of the classical IPM does not depend on κ and its dependence on ϵ^{-1} is logarithmic. Dependence on the condition number bound κ (which in the overall running time expression for the algorithm of [ANTZ21] is quadratic) is particularly problematic in the context of IPMs, as the condition number of the Newton linear system almost always goes to infinity as we approach optimality.

5.1.2   Our results

In this chapter, we study an Inexact-Feasible QIPM for SDO, and how iterative refining techniques can be leveraged to obtain exponentially better running times than QIPMs presently found in the literature.

First, we demonstrate that the IR methodology for classically solving linear systems discussed in Chapter 4 as a quantum subroutine for solving the Newton linear system in each iteration. This leads to an exponential $\mathcal{O}\left(\frac{1}{\epsilon}\right)$ speedup in the complexity of the IF-QIPM from [ANTZ21], and ensures that the algorithm's dependence on the inverse of the precision to which we solve the SDO problem at hand will be polylogarithmic. This leads to an IF-QIPM that converges to an ϵ-optimal solution to an SDO problem involving $n \times n$ matrices and $m = \mathcal{O}(n^2)$ constraints using at most

$$\mathcal{O}\left(n^{3.5}\kappa^2 \cdot \mathrm{polylog}\left(n, \kappa, \frac{1}{\epsilon}\right)\right)$$

QRAM accesses and $\mathcal{O}\left(n^{4.5} \cdot \mathrm{polylog}\left(n, \kappa, \frac{1}{\epsilon}\right)\right)$ arithmetic operations. A de-quantization of our approach leads to a novel classical IF-IPM, which converges to an ϵ-optimal solution in time

$$\mathcal{O}\left(n^{4.5}\kappa \cdot \mathrm{polylog}\left(n, \kappa, \frac{1}{\epsilon}\right)\right).$$

5.2 Primal and Dual SDOs and the Central Path

We denote the feasible sets of (5.1) and (5.2) by

$$\mathcal{P} = \left\{ X \in \mathcal{S}^n : \operatorname{tr}(A_i X) = b_i, \ i \in [m], X \succeq 0 \right\},$$

$$\mathcal{D} = \left\{ (y, S) \in \mathbb{R}^m \times \mathcal{S}^n : \sum_{i=1}^{m} y_i A_i + S = C, S \succeq 0 \right\}.$$

For our work here, we assume that a strictly feasible pair X and (y, S) with $X \succ 0$ and $S \succ 0$ exists, i.e., the IPC is satisfied [dKRT97]. This may be assumed without loss of generality, one could cast the original SDO problem as a slightly larger one using the self-dual embedding model, for which a strictly feasible initial solution always exists [dKRT97]. In particular, the analytic center of our space, i.e., $X^{(0)} = S^{(0)} = n^{-1}I$ is always strictly feasible in this setting.

The sets of *interior feasible solutions* are defined by

$$\mathcal{P}^0 = \left\{ X \in \mathcal{P} : X \succ 0 \right\},$$

$$\mathcal{D}^0 = \left\{ (y, S) \in \mathcal{D} : S \succ 0 \right\}.$$

Under the IPC, it is guaranteed that the primal and dual optimal sets:

$$\mathcal{P}^* = \left\{ X \in \mathcal{P} : \operatorname{tr}(CX) = z_P \right\},$$

$$\mathcal{D}^* = \left\{ (y, S) \in \mathcal{D} : b^\top y = z_D \right\},$$

are nonempty and bounded, with $z_P = z_D$,. That is, for all optimal solutions X^* and (y^*, S^*), the duality gap is zero, with

$$\operatorname{tr}(CX^*) - b^\top y^* = \operatorname{tr}(X^* S^*) = 0. \tag{5.3}$$

As a consequence, $X^* S^* = S^* X^* = 0$ as X^* and S^* are symmetric positive semidefinite matrices. However, for primal-dual IPMs, the complementarity condition $XS = 0$ is perturbed to:

$$XS = \sigma\mu I, \tag{5.4}$$

where I is the $n \times n$ identity matrix, $\sigma \in [0, 1]$ is the centering parameter, and $\mu > 0$ is the so-called central path parameter, which is progressively reduced to zero in the course of the optimization algorithm.

For all $\mu > 0$, assuming the IPC and linear independence of the matrices A_i for $i \in [m]$, the central path equation system

$$\text{tr}\,(A_i X) = b_i \ \forall i \in [m], \ X \succ 0,$$

$$\sum_{i \in [m]} y_i A_i + S = C, \ S \succ 0, \tag{5.5}$$

$$XS = \sigma\mu I,$$

has a unique solution [NN88]. The set of solutions for all $\mu > 0$ gives the central path for the primal and dual SDO problems. In IPMs we aim to follow the central path as $\mu \to 0$, i.e., as we approach optimality. In fact, we simply seek to stay in a certain neighborhood of the central path.

A classical IPM, as outlined in Algorithm 3, begins with a strictly feasible primal-dual pair $\left(X^{(0)}, S^{(0)}\right) \in \mathcal{P}^0 \times \mathcal{D}^0$ for the primal and dual SDO problems (5.1) and (5.2). This pair has a duality gap of $\text{tr}\left(X^{(0)} S^{(0)}\right) = \mu^{(0)} n$, and a distance to the central path of $d\left(X^{(0)}, S^{(0)}\right) \le \gamma\mu^{(0)}$ for $\gamma \in (0,1)$, where $\mu_0 = \dfrac{\text{tr}\left(X^{(0)} S^{(0)}\right)}{n}$, for some appropriate distance metric $d(X,S)$. Without loss of generality, one can assume that $\mu^{(0)} = \mathcal{O}(1)$. As we mentioned earlier, one it is always possible obtain an interior feasible solution using the self-dual embedding model, and in particular, the analytic center of our space, i.e., $X^{(0)} = S^{(0)} = n^{-1} I$ is always strictly feasible in this setting.

In this work we consider the narrow (or, Frobenius) neighborhood

$$\mathcal{N}_F(\gamma) = \left\{ (X, y, S) \in \mathcal{P}^0 \times \mathcal{D}^0 : \left\| X^{1/2} S X^{1/2} - \mu I \right\|_F = \left[\sum_{i=1}^{n} (\lambda_i(XS) - \mu)^2 \right]^{1/2} \le \gamma\mu \right\}.$$

Another popular alternative is the so called *negative infinity neighborhood* that is a *large neighborhood*, defined as

$$\mathcal{N}_\infty^-(1 - \gamma) = \{(X, y, S) \in \mathcal{P}^0 \times \mathcal{D}^0 : \lambda_{\min}(XS) \ge \gamma\mu\},$$

where $\gamma \in (0, 1)$. It is well known that classical feasible primal-dual path following IPMs that use the former neighborhood exhibit an iteration complexity of the order $\mathcal{O}(\sqrt{n}\log(1/\epsilon))$ whereas long-step IPMs, using the latter neighborhood, have iteration complexity $\mathcal{O}(n\log(1/\epsilon))$. However, in practice the long step algorithms tend to converge faster than the short step counterpart; in this paper we focus on the theoretical running time and therefore use the narrow (or Frobenius) neighborhood $\mathcal{N}_F(\gamma)$. Thus, we define our centrality measure at a point $(X, S) \in \mathcal{S}_+^n \times \mathcal{S}_+^n$, with $\mu = \frac{\text{tr}(XS)}{n}$, to be

$$d(X, S) = \left\| X^{1/2} S X^{1/2} - \mu I \right\|_F. \tag{5.6}$$

Then, in each iteration, we solve the so-called Newton linear system in order to obtain ΔX

and ΔS and update the solutions using the following rule:

$$X = X + \Delta X,$$
$$S = S + \Delta S.$$

If we were to directly linearize the complementarity condition, the Newton linear system would then be written as:

$$\Delta X S + \Delta S X = \sigma \mu I - X S,$$
$$\Delta X \in \text{Null}(A_1, A_2, \ldots, A_m), \tag{5.7}$$
$$\Delta S \in \mathcal{R}(A_1, A_2, \ldots, A_m),$$

where $\text{Null}(\cdot), \mathcal{R}(\cdot)$ are the nullspace (or, *kernel*) and rowspace of the constraint matrices, respectively (we provide explicit definitions for these subspaces in the following section). However, it is known that (5.7) does not have a symmetric matrix solution with respect to ΔX, see, e.g., [AHO98]. This must be addressed, because both the primal and dual solutions must be symmetric.

Algorithm 3 Classical interior point method

Input: $\epsilon, \delta > 0$; $\sigma = 1 - \delta/\sqrt{n}$; $\gamma \in (0, 1)$
Choose $(X^{(0)}, y^{(0)}, S^{(0)}) \in \mathcal{N}_F(\gamma)$
Choose a nonsingular matrix P from Monteiro-Zhang family of search directions, compute $P^{(0)}$
Set $\mu^{(0)} \leftarrow \dfrac{\text{tr}\left(X^{(0)} S^{(0)}\right)}{n}$, $k \leftarrow 0$
while $\mu > \epsilon$ **do**

    1. $\mu^{(k)} \leftarrow \dfrac{\text{tr}\left(X^{(k)} S^{(k)}\right)}{n}$

    2. Compute the scaling matrix $P^{(k)}$

    3. Compute the solution $\left(\Delta X^{(k)}, \Delta y^{(k)}, \Delta S^{(k)}\right)$ to the Newton linear system (5.10) defined by choice of $P^{(k)}$

$$X^{(k+1)} \leftarrow X^{(k)} + \Delta X^{(k)}, \ S^{(k+1)} \leftarrow S^{(k)} + \Delta S^{(k)} \text{ and } y^{(k+1)} \leftarrow y^{(k)} + \Delta y^{(k)}$$
$$k \leftarrow k + 1$$

end

5.2.1 Symmetrizing the Newton System

As has been extensively studied in the literature of IPMs (see, e.g., [AHO98, NT97, NT98]), the Newton steps ΔS and ΔX have to be symmetric matrices. More specifically, even though ΔS is guaranteed to be symmetric by the definition of a dual feasible solution, there is no symmetric ΔX that solves system (5.7). Thus, we need a different approach to guarantee symmetry.

In an effort to generalize the scaling methods required for primal-dual symmetry, following Zhang [Zha98], we define the linear transformation $H_P(M)$ that symmetrizes a matrix M using a given invertible matrix P:

$$H_P(M) = \frac{1}{2}\left[PMP^{-1} + P^{-\top}M^\top P^\top\right]. \tag{5.8}$$

Observe that $H_P(M) = \sigma\mu I \iff M = \sigma\mu I$; we can then write the central path equations in symmetric form as: $H_P(XS) = \sigma\mu I$. In this light, we can also symmetrize the linearized complementarity condition of the Newton system $X\Delta S + S\Delta X = \sigma\mu I - XS$ as

$$H_P(X\Delta S + S\Delta X) = \sigma\mu I - H_P(XS). \tag{5.9}$$

Equation (5.9) can be explicitly expressed as:

$$P(X\Delta S + \Delta XS)P^{-1} + P^{-\top}(\Delta SX + S\Delta X)P^\top = 2\sigma\mu I - PXSP^{-1} - P^{-\top}SXP^\top.$$

Defining $\mathcal{A}^\top = [\mathrm{vec}(A_1)\cdots\mathrm{vec}(A_m)]$, we can write the symmetrized Newton linear system as

$$\begin{pmatrix} 0 & \mathcal{A}^\top & I_{n^2} \\ \mathcal{A} & 0 & 0 \\ \mathcal{E} & 0 & \mathcal{F} \end{pmatrix}\begin{pmatrix} \mathrm{vec}(\Delta X) \\ \Delta y \\ \mathrm{vec}(\Delta S) \end{pmatrix} = \begin{pmatrix} 0 \\ 0 \\ \mathrm{vec}(R^c) \end{pmatrix},$$

where I_{n^2} is the identity of order n^2 and

$$\mathcal{E} = P \otimes SP^{-1} + P^{-1}S \otimes P,$$
$$\mathcal{F} = PX \otimes P^{-1} + P^{-1} \otimes XP,$$
$$R^c = \sigma\mu I - H_P(XS).$$

Now let us define $\mathcal{A}_s^\top = [\mathrm{svec}(A_1)\cdots\mathrm{svec}(A_m)]$. Following Todd et al. [TTT98], we can also write the Newton linear system using the svec notation, in which case we have:

$$\begin{pmatrix} 0 & \mathcal{A}_s & \mathcal{I} \\ \mathcal{A}_s^\top & 0 & 0 \\ \mathcal{E}_s & 0 & \mathcal{F}_s \end{pmatrix}\begin{pmatrix} \mathrm{svec}(\Delta X) \\ \Delta y \\ \mathrm{svec}(\Delta S) \end{pmatrix} = \begin{pmatrix} 0 \\ 0 \\ R_s^c \end{pmatrix}, \tag{5.10}$$

where $\mathcal{I}$ is the identity matrix of order $\frac{n(n+1)}{2}$ and

$$\mathcal{E}_s = P \otimes_s P^{-\top}S,$$
$$\mathcal{F}_s = PX \otimes_s P^{-\top},$$
$$R_s^c = \mathrm{svec}(\sigma\mu I - H_P(XS)) = \mathrm{svec}(R^c).$$

One can note that matrices $\mathcal{E}_s$ and $\mathcal{F}_s$ are nonsingular whenever X and S are positive definite matrices [TTT98]. We have:

$$\mathcal{E}_s = P \otimes_s P^{-\top} S = (I \otimes_s P^{-\top} S P^{-1})(P \otimes_s P), \tag{5.11a}$$

$$\mathcal{F}_s = P X \otimes_s P^{-\top} = (P X P^{\top} \otimes_s I)(P^{-\top} \otimes_s P^{-\top}), \tag{5.11b}$$

where the equalities follow from the definition of the symmetric Kronecker product detailed earlier.

We then need to choose P to obtain specific instances of this system, and we refer to the class of search directions parameterised by P as the Monteiro-Zhang family of search directions [MZ98]. We present results involving three possible choice of P: $P = I$, the AHO direction [AHO98]; $P = S^{1/2}$, the HKM direction [HRVW96, KSH97, Mon98]; and $P = W^{-1/2}$, where W is the Nesterov-Todd [NT98] scaling matrix, defined as:

$$W = S^{-1/2}(S^{1/2} X S^{1/2})^{1/2} S^{-1/2} = X^{1/2}(X^{1/2} S X^{1/2})^{-1/2} X^{1/2}. \tag{5.12}$$

Letting (X, y, S) be the current solution at iteration k, we can define the HKM Newton linear system for the k-th iteration by using symmetry to drop all of the transpose terms involving P and set:

$$\mathcal{E}_s = S^{1/2} \otimes_s S^{-1/2} S = S^{1/2} \otimes_s S^{1/2}, \tag{5.13a}$$

$$\mathcal{F}_s = S^{1/2} X \otimes_s S^{-1/2}, \tag{5.13b}$$

$$R_s^c = \operatorname{svec}(2\sigma\mu I - S^{1/2} X S^{1/2}). \tag{5.13c}$$

In defining the AHO Newton Linear system, taking $P = I$ simplifies the above quantities to to

$$\mathcal{E}_s = I \otimes_s S,$$

$$\mathcal{F}_s = X \otimes_s I,$$

$$R_s^c = \operatorname{svec}(2\sigma\mu I - XS - SX).$$

Although this guarantees that our solutions to the Newton system are symmetric, the AHO direction only guarantees a solution if we are within a small (local) neighborhood of the central path: the ∞-norm neighborhood with opening less than $1/3$. Finally, for the Nesterov-Todd Newton system, we set $P = W^{-1/2}$ and we can write:

$$\mathcal{E}_s = W^{-1/2} \otimes_s W^{1/2} S, \tag{5.14a}$$

$$\mathcal{F}_s = W^{-1/2} X \otimes_s W^{1/2}, \tag{5.14b}$$

$$R_s^c = \operatorname{svec}\left(2\sigma\mu I - W^{-1/2} X S W^{1/2} - W^{1/2} S X W^{-1/2}\right). \tag{5.14c}$$

Symmetry is not the only issue overlooked by previous works on QIPMs; we still need to account for the consequences of inexact tomography in the context of QIPMs.

5.3   Newton Linear Systems for QIPMs

We now turn our attention to solving the Newton linear system using quantum linear system solvers. In particular, we present the methodology introduced in [ANTZ21], which allows one to design an inexact, but feasible primal-dual QIPM framework which yields a speedup in n over the classical variants.

5.3.1   The Nullspace Representation of the Newton Linear System

Analogous to Gondzio's [Gon13] IF-IPM for convex quadratic optimization over linear constraints, in our novel IF-QIPM approach the first two equations of (5.10) will hold, while we assume that the third equation is only satisfied up to some residual error $R^r \in \mathcal{S}^n$, which in our setting is due to the use of quantum state tomography (subsequent discussion will show that we can guarantee the residual is symmetric):

$$\begin{pmatrix} 0 & \mathcal{A}_s^\top & \mathcal{I} \\ \mathcal{A}_s & 0 & 0 \\ \mathcal{E}_s & 0 & \mathcal{F}_s \end{pmatrix} \begin{pmatrix} \mathrm{svec}(\Delta X) \\ \Delta y \\ \mathrm{svec}(\Delta S) \end{pmatrix} = \begin{pmatrix} 0 \\ 0 \\ R_s^c + \mathrm{svec}(R^r) \end{pmatrix}, \tag{5.15}$$

where $R_s^c = \mathrm{svec}(\sigma\mu I - H_P(XS)) = \mathrm{svec}(R^c)$. As in [Gon13], we assume that the residual term R^r, for some $\beta \in (0,1)$, satisfies

$$\|R^r\|_F \le \beta \|R^c\|_F. \tag{AR1}$$

In this way, we ensure that the residual norms are being driven towards zero as we approach optimality, due to the fact that $\|R^c\|_F \to 0$ as $\mu \to 0$. Additionally, choosing the level of error in this manner is not very restrictive, as the algorithm terminates as soon as we reach a solution satisfying $\mu \le \epsilon$.

The ensuing result from [TTT98] establishes the uniqueness of the solution to the system (5.15).

Theorem 5.1 (Theorem 3.1 in [TTT98]). *Suppose X and S are positive definite. Then the system of equations (5.15) has a unique solution $(\Delta X, \Delta y, \Delta S) \in \mathcal{S}^n \times \mathbb{R}^m \times \mathcal{S}^n$ if $\mathcal{E}_s^{-1}\mathcal{F}_s$ is positive definite (not necessarily symmetric). In particular, this holds when X, S and $H_P(XS)$ are positive definite.*

Proof. The proof is the same as the one given in [TTT98], but we repeat it here for completness. In what follows, positive definiteness does not imply symmetry. We seek to show that the system

$$\begin{pmatrix} 0 & \mathcal{A}_s & 0 \\ \mathcal{A}_s^\top & 0 & \mathcal{I} \\ 0 & \mathcal{E}_s & \mathcal{F}_s \end{pmatrix} \begin{pmatrix} \Delta y \\ \mathrm{svec}(\Delta X) \\ \mathrm{svec}(\Delta S) \end{pmatrix} = 0 \tag{5.16}$$

is satisfied only by the all-zero solution. Note that for convenience, we have reordered the block-rows of the Newton system in order to utilize block-Gaussian elimination for the purposes of this proof. From (5.11a) it follows that $\mathcal{E}_s$ is invertible. Hence, applying block-Gaussian elimination to (5.16) yields the Schur complement equation:

$$(\mathcal{A}_s \mathcal{E}_s^{-1} \mathcal{F}_s \mathcal{A}_s^\top)\Delta y = 0.$$

By assumption, $\mathcal{E}_s^{-1}\mathcal{F}_s \succ 0$ and $\mathcal{A}_s$ has full row rank, implying that the matrix $\mathcal{A}_s \mathcal{E}_s^{-1}\mathcal{F}_s\mathcal{A}_s^\top$ is positive definite, therefore $\Delta y = 0$. Thus the second equation in (5.16) gives $\Delta S = -\operatorname{smat}(\mathcal{A}_s^\top \Delta y) = 0$ and from the third equation it follows that $\Delta X = -\mathcal{E}_s^{-1}\mathcal{F}_s\Delta S = 0$. This completes the proof. $\qquad\square$

If the Newton linear system as given by (5.15) is solved using QLSA and tomography, the first two block equations in this system will not be satisfied exactly, either. To design our IF-QIPM, we re-write the optimality conditions (5.10) as:

$$\Delta X \in \operatorname{Null}(\mathcal{A}_s)$$
$$\Delta S \in \mathcal{R}(\mathcal{A}_s) \tag{5.17}$$
$$(P \otimes_s P^{-\top}S)\operatorname{svec}(\Delta X) + (PX \otimes_s P^{-\top})\operatorname{svec}(\Delta S) = \operatorname{svec}(\sigma\mu I - H_P(XS))$$

where $\operatorname{Null}(\mathcal{A}_s)$ denotes the nullspace of $\mathcal{A}_s$ and $\mathcal{R}(\mathcal{A}_s)$ denotes the rowspace of $\mathcal{A}_s$. Crucially, $\operatorname{Null}(\mathcal{A}_s)$ and $\mathcal{R}(\mathcal{A}_s)$ are orthogonal subspaces to one another; we always have $\operatorname{tr}(\Delta X \Delta S) = 0$ for any $\Delta X \in \operatorname{Null}(\mathcal{A}_s)$ and $\Delta S \in \mathcal{R}(\mathcal{A}_s)$.

For $\mathcal{R}(\mathcal{A}_s)$, we choose

$$\{\operatorname{svec}(A_1) \cdots \operatorname{svec}(A_m)\}$$

as a basis. If the primal-dual pair (5.1)-(5.2) is given in canonical form (inequalities instead of equalities), we can trivially obtain basses for the nullspace and rowspace directly from the coefficient matrix. In this case, one would use the Self-Dual embedding model for that form, the nullspace basis matrix can be constructed trivially too from the coefficient matrix (without needing factorization or Gaussian elimination). We can also calculate a basis of $\operatorname{Null}(\mathcal{A}_s)$ using either Gaussian elimination or the QR-Factorization. Recall that $\mathcal{A}_s \in \mathbb{R}^{m \times \frac{n(n+1)}{2}}$, so $\mathcal{A}_s^\top \in \mathbb{R}^{\frac{n(n+1)}{2} \times m}$. The QR factorization of $\mathcal{A}_s^\top$ is defined as:

$$\mathcal{A}_s^\top = \begin{bmatrix} Q_1 & Q_2 \end{bmatrix} \begin{bmatrix} R \\ 0 \end{bmatrix},$$

where $Q_1 \in \mathbb{R}^{\frac{n(n+1)}{2} \times m}$ and $Q_2 \in \mathbb{R}^{\frac{n(n+1)}{2} \times \left(\frac{n(n+1)}{2} - m\right)}$. Then, it follows that the columns of Q_2 form a basis for the null space of $\mathcal{A}_s$. Hence, introducing a variable $\Delta z \in \mathbb{R}^{\left(\frac{n(n+1)}{2} - m\right)}$, ΔX and

ΔS can be written as

$$\mathrm{svec}(\Delta X) = Q_2 \Delta z, \tag{5.18a}$$

$$\mathrm{svec}(\Delta S) = -\mathcal{A}_s^\top \Delta y. \tag{5.18b}$$

The following result defines the Newton linear system to be solved at each iteration of the IF-QIPM.

Proposition 5.2. *Let P be a nonsingular scaling matrix from the Monteiro-Zhang family of search directions, and define*

$$\mathcal{E}_s = (P \otimes_s P^{-\top} S),$$

$$\mathcal{F}_s = (PX \otimes_s P^{-\top}),$$

$$R_s^c = \mathrm{svec}(\sigma \mu I - H_P(XS)).$$

The Newton linear system for the IF-QIPM is given by

$$\begin{bmatrix} \mathcal{E}_s Q_2 & \mathcal{F}_s(-\mathcal{A}_s^\top) \end{bmatrix} \begin{bmatrix} \Delta z \\ \Delta y \end{bmatrix} = R_s^c. \tag{5.19}$$

Proof. Substituting equations (5.18a) and (5.18b) into the left-hand side of the third equation in system (5.17) yields

$$(P \otimes_s P^{-\top} S)\,\mathrm{svec}(\Delta X) + (PX \otimes_s P^{-\top})\,\mathrm{svec}(\Delta S)$$

$$= (P \otimes_s P^{-\top} S)(Q_2 \Delta z) + (PX \otimes_s P^{-\top})(-\mathcal{A}_s^\top \Delta y)$$

$$= \begin{bmatrix} \mathcal{E}_s Q_2 & \mathcal{F}_s(-\mathcal{A}_s^\top) \end{bmatrix} \begin{bmatrix} \Delta z \\ \Delta y \end{bmatrix}.$$

$\square$

Observe that the coefficient matrix of equation (5.19) is full rank, as the columns of $A_s^\top$ and Q_2 are orthogonal, *and* the columns of $A_s^\top$ and Q_2 are linearly independent. Thus, equation (5.19) has a unique solution. Upon using QLSA to solve the system (5.19) for $|\Delta z \circ \Delta y)$, we will need to map classical estimates $(\overline{\Delta z}, \overline{\Delta y})$ of the solution to a classical estimate $(\overline{\Delta X}, \overline{\Delta y}, \overline{\Delta S})$ in order to proceed to the next iteration. Define ΔX and ΔS by (5.18a) and (5.18b), respectively. As $(\Delta z, \Delta y)$ is the unique solution of (5.19), consequently ΔX and ΔS are uniquely determined as well.

Our use of inexact tomography causes Δz and Δy to be calculated with some error. Yet, by using the bases of the nullspace and the rowspace to calculate ΔX and ΔS, we ensure that these search directions are always coming from orthogonal subspaces, regardless of the errors introduced by tomography or the length of the stepsize. We formalize this result in the following proposition.

Proposition 5.3. *For a given $X \in \mathcal{P}^0$ and $(y, S) \in \mathcal{D}^0$, let $(\overline{\Delta z}, \overline{\Delta y})$ be an estimate of the solution to (5.19). Then,*

$$\overline{\Delta z} = \Delta z + \xi_z \ \text{ and } \ \overline{\Delta y} = \Delta y + \xi_y,$$

where ξ_z and ξ_y denote the error terms of system (5.19). Let $\overline{\Delta X}$ and $\overline{\Delta S}$ be given by (5.18a) and (5.18b), respectively. Then for any step size α, we have

$$\mathcal{A}_s(\operatorname{svec}(X) + \alpha \operatorname{svec}(\overline{\Delta X})) = b, \ \text{ and } \ \mathcal{A}_s^\top(y + \alpha \overline{\Delta y}) + (\operatorname{svec}(S) + \alpha \operatorname{svec}(\overline{\Delta S})) = \operatorname{svec}(C).$$

Proof. First, note that for any $X \in \mathcal{P}$ and $(y, S) \in \mathcal{D}$, we have

$$\mathcal{A}_s \operatorname{svec}(X) = b$$
$$\mathcal{A}_s^\top y + \operatorname{svec}(S) = \operatorname{svec}(C).$$

Therefore,

$$
\begin{aligned}
\mathcal{A}_s(\operatorname{svec}(X) + \alpha \operatorname{svec}(\overline{\Delta X})) &= \mathcal{A}_s \operatorname{svec}(X) + \alpha \mathcal{A}_s \operatorname{svec}(\overline{\Delta X}) \\
&= b + \alpha \mathcal{A}_s [Q_2(\overline{\Delta z})] \\
&= b + \alpha [\mathcal{A}_s Q_2(\Delta z + \xi_z)] \\
&= b + \alpha [0 \cdot (\Delta z + \xi_z)] \\
&= b.
\end{aligned}
$$

Similarly,

$$
\begin{aligned}
\mathcal{A}_s^\top(y + \alpha \overline{\Delta y}) + (\operatorname{svec}(S) + \alpha \operatorname{svec}(\overline{\Delta S})) &= [\mathcal{A}_s^\top y + \operatorname{svec}(S)] + [\mathcal{A}_s^\top \alpha \overline{\Delta y} + \alpha \operatorname{svec}(\overline{\Delta S})] \\
&= \operatorname{svec}(C) + \alpha [\mathcal{A}_s^\top \overline{\Delta y} + \operatorname{svec}(\overline{\Delta S})] \\
&= \operatorname{svec}(C) + \alpha [\mathcal{A}_s^\top (\overline{\Delta y} - \overline{\Delta y})] \\
&= \operatorname{svec}(C).
\end{aligned}
$$

$\square$

Having established the fact that our search directions computed from inexact estimates of the solution to (5.19) preserve primal and dual feasibility, we next certify the validity of using the system (5.19) to obtain a solution to the system (5.15). Further, we demonstrate that if the system (5.15) has a unique solution, than the system (5.19) does as well. The proof of the next result is straightforward using the definition of the quantities involved.

Proposition 5.4. *Suppose X and S are positive definite. Then, any solution $(\Delta z, \Delta y)$ of the system (5.19), provides a solution $(\Delta X, \Delta y, \Delta S)$ to the system (5.15).*

The following result from [Mon98] plays a crucial role in the analysis required to establish polynomial convergence of our IF-QIPM.

Lemma 5.5 (Lemma 2.1 in [Mon98]). *Suppose that $(X, S) \in \mathcal{S}_{++}^n \times \mathcal{S}_{++}^n$, $Q \in \mathbb{R}^{n \times n}$ is a nonsingular matrix and $\mu = \frac{\operatorname{tr}(XS)}{n}$. Then:*

(a) $d(\widetilde{X}, \widetilde{S}) = d(X, S)$ where $\widetilde{X} = QXQ^\top$ and $\widetilde{S} = Q^{-\top}SQ^{-1}$.

(b) $d(X, S) \leq \|H_Q(XS - \mu I)\|_F$ with equality holding if $QXSQ^{-1} \in \mathcal{S}^n$.

5.4   Technical Results

We now present the technical results required to prove our IF-QIPM for SDO exhibits a polynomial iteration complexity. Our convergence analysis can be viewed as a generalization of Gondzio's [Gon13] analysis from LO to SDO. To accomplish this, we adapt the work of Monteiro [Mon98] to account for the inexactness of the complementarity condition in order to prove polynomial convergence of the IF-QIPM presented in Section 5.5. Though in this work we chose the Nesterov-Todd direction $P = W^{-1/2}$, the convergence analysis we perform is presented in general terms of P, and holds for any member of the Monteiro-Zhang family of search directions [MZ98], which includes the NT, AHO and HKM directions.

Following [Mon98], we assume that $(X, y, S) \in \mathcal{S}_{++}^n \times \mathbb{R}^m \times \mathcal{S}_{++}^n$ and that $P \in \mathbb{R}^{n \times n}$ is a nonsingular matrix. Additionally, we assume $(\Delta X, \Delta y, \Delta S)$ is a solution of the system

$$\operatorname{tr}(A_i \Delta X) = b_i - \operatorname{tr}(A_i X), \quad \forall i \in [m], \tag{5.20a}$$

$$\sum_{i=1}^m \Delta y_i A_i + \Delta S = C - S - \sum_{i=1}^m y_i A_i, \tag{5.20b}$$

$$H_P(\Delta X S + X \Delta S) = \sigma \mu I - H_P(XS) + R^r, \tag{5.20c}$$

for some $\sigma \in [0, 1]$. Recall that $(\Delta X, \Delta y, \Delta S)$ is a solution to the system (5.15) obtained from our classical estimate $(\Delta z, \Delta y)$ of the solution $|\Delta z \circ \Delta y\rangle$ to the Newton linear system (5.19). That is,

$$(\Delta X, \Delta y, \Delta S) = \left(\operatorname{smat}(Q_2 \Delta z), \Delta y, -\operatorname{smat}(\mathcal{A}_s^\top \Delta y)\right), \tag{5.21}$$

where ΔX and ΔS are computed according to (5.18a) and (5.18b), respectively. Further, for

$\alpha \in \mathbb{R}$ we define the quantities:

$$\tilde{X} = PXP^{\top}, \qquad\qquad\qquad\qquad\qquad \tilde{S} = P^{-\top}SP^{-1}, \qquad (5.22\text{a})$$

$$\widetilde{\Delta X} = P\Delta X P^{\top}, \qquad\qquad\qquad\qquad \widetilde{\Delta S} = P^{-\top}\Delta S P^{-1}, \qquad (5.22\text{b})$$

$$X(\alpha) = X + \alpha\Delta X, \qquad y(\alpha) = y + \alpha\Delta y, \qquad S(\alpha) = S + \alpha\Delta S, \qquad (5.22\text{c})$$

$$\tilde{X}(\alpha) = PX(\alpha)P^{\top} = \tilde{X} + \alpha\widetilde{\Delta X}, \qquad (5.22\text{d})$$

$$\tilde{S}(\alpha) = P^{-\top}S(\alpha)P^{-1} = \tilde{S} + \alpha\widetilde{\Delta S}, \qquad (5.22\text{e})$$

$$\mathcal{W}_X = \tilde{X}^{-1/2}\left[\widetilde{\Delta X}\tilde{S} + \tilde{X}\widetilde{\Delta S} + \tilde{X}\tilde{S} - \sigma\mu I - \tilde{R}^r\right]\tilde{X}^{1/2}, \qquad (5.22\text{f})$$

and

$$\mu = \frac{\operatorname{tr}(XS)}{n} = \frac{\operatorname{tr}\left(\tilde{X}\tilde{S}\right)}{n}, \quad \mu(\alpha) = \frac{\operatorname{tr}\left(X(\alpha)S(\alpha)\right)}{n} = \frac{\operatorname{tr}\left(\tilde{X}(\alpha)\tilde{S}(\alpha)\right)}{n}. \qquad (5.23)$$

Lemma 5.6 (Lemma 3.1 in [Mon98]). *$(\Delta X, \Delta y, \Delta S)$ is a solution of the system (5.20) if and only if $(\widetilde{\Delta X}, \Delta y, \widetilde{\Delta S})$ is a solution of the system*

$$\operatorname{tr}\left(\tilde{A}_i\widetilde{\Delta X}\right) = b_i - \operatorname{tr}\left(\tilde{A}_i\tilde{X}\right), \quad \forall i \in [m], \qquad (5.24\text{a})$$

$$\sum_{i=1}^{m}\Delta y_i \tilde{A}_i + \widetilde{\Delta S} = \tilde{C} - \tilde{S} - \sum_{i=1}^{m} y_i \tilde{A}_i \qquad (5.24\text{b})$$

$$H_I(\widetilde{\Delta X}\tilde{S} + \tilde{X}\widetilde{\Delta S}) = \sigma\mu I - H_I(\tilde{X}\tilde{S}) + \tilde{R}^r \qquad (5.24\text{c})$$

where $\tilde{R}^r = P^{-\top}R^r P^{-1}$, $\tilde{C} = P^{-\top}CP^{-1}$ and $\tilde{A}_i = P^{-\top}\tilde{A}_i P^{-1}$ for $i \in [m]$.

Proof. The proof follows as a consequence of equations (5.20), (5.22a) and (5.22b). $\qquad\square$

In what follows, we work under the assumption that

$$(X, y, S) \in \mathcal{P}^0 \times \mathcal{D}^0.$$

We remark that in the scaled setting our assumption for the residual term in (AR1) is given by:

$$\|\tilde{R}^r\|_F \leq \beta\|\tilde{R}^c\|_F, \qquad (\text{AR2})$$

where $\beta \in (0,1)$ and $\tilde{R}^c = \sigma\mu I - H_I(\tilde{X}\tilde{S})$. For the analysis in this section we assume that the error tolerances are chosen so that the residual term $\tilde{R}^r$ for the scaled problem (5.24) in each iteration satisfies

$$\|\tilde{X}^{-1/2}\tilde{R}^r\tilde{X}^{1/2}\|_F \leq \beta\|\tilde{X}^{-1/2}\tilde{R}^c\tilde{X}^{1/2}\|_F \qquad (\text{AR3})$$

for some $\beta \in (0,1)$. Note that these assumptions we make regarding the residual are mild. In light of Lemma 5.5(a), R^r and $\tilde{R}^r$ satisfy the same Frobenius norm bound, and the right hand sides of the inequalities in (AR1), (AR2) and (AR3) are all invariant under the stated scaling.

Lemma 5.7 (Lemma 3.3 in [Mon98]). *The following relations hold:*

$$H_{\widetilde{X}^{1/2}}(\mathcal{W}_X) = 0, \tag{5.25}$$

$$\operatorname{tr}\left(\widetilde{\Delta X \Delta S}\right) = \operatorname{tr}\left(\Delta X \Delta S\right) = 0. \tag{5.26}$$

Proof. Equation (5.25) is a direct consequence of (5.24c). Further from our assumption

$$(X, y, S) \in \mathcal{P}^0 \times \mathcal{D}^0,$$

and applying equations (5.20c), (5.20b) and (5.22b) yields the two identities in (5.26). $\qquad\square$

The following result can be viewed as the inexact analogue of Lemma 3.4 in [Mon98].

Lemma 5.8. *For all $\alpha \in \mathbb{R}$ we have*

$$\mu(\alpha) = (1 - \alpha + \sigma\alpha)\mu + \alpha\frac{\operatorname{tr}(R^r)}{n}, \tag{5.27}$$

$$\widetilde{X}^{-1/2}\left[\widetilde{X}(\alpha)\widetilde{S}(\alpha) - \mu(\alpha)I\right]\widetilde{X}^{1/2} = (1 - \alpha)(\widetilde{X}^{1/2}\widetilde{S}\widetilde{X}^{1/2} - \mu I) + \alpha\mathcal{W}_X + \alpha^2\widetilde{X}^{-1/2}\widetilde{\Delta X \Delta S}\widetilde{X}^{1/2}$$

$$+ \alpha\widetilde{X}^{-1/2}\left[\widetilde{R}^r - \frac{\operatorname{tr}(R^r)}{n}I\right]\widetilde{X}^{1/2}. \tag{5.28}$$

Proof. From (5.22c) we have

$$X(\alpha)S(\alpha) = (X + \alpha\Delta X)(S + \alpha\Delta S) = XS + \alpha(X\Delta S + \Delta X S) + \alpha^2\Delta X \Delta S.$$

Hence, by the linearity of $H_P(\cdot)$ and equation (5.20c), it follows

$$X(\alpha)S(\alpha) = (X + \alpha\Delta X)(S + \alpha\Delta S) = H_P(XS) + \alpha H_P(X\Delta S + \Delta X S) + \alpha^2 H_P(\Delta X \Delta S)$$

$$= H_P(XS) + \alpha[H_P(\sigma\mu I - XS) + R^r] + \alpha^2 H_P(\Delta X \Delta S)$$

$$= (1 - \alpha)H_P(XS) + \alpha\sigma\mu I + \alpha R^r + \alpha^2 H_P(\Delta X \Delta S).$$

Now, for $M \in \mathbb{R}^{n \times n}$ we have[1] $\operatorname{tr}(H_P(M)) = \operatorname{tr}(M)$, and thus

$$\operatorname{tr}\left(X(\alpha)S(\alpha)\right) = \operatorname{tr}[X(\alpha)S(\alpha)] = \operatorname{tr}[H_P(X(\alpha)S(\alpha))]$$

$$= \operatorname{tr}[(1 - \alpha)H_P(XS) + \alpha\sigma\mu I + \alpha R^r + \alpha^2 H_P(\Delta X \Delta S)]$$

$$= (1 - \alpha)\operatorname{tr}[H_P(XS)] + \alpha\sigma\mu\operatorname{tr}(I) + \alpha\operatorname{tr}(R^r) + \alpha^2\operatorname{tr}[H_P(\Delta X \Delta S)]$$

$$= (1 - \alpha)\operatorname{tr}(XS) + \alpha\sigma\mu n + \alpha\operatorname{tr}(R^r) + \alpha^2\operatorname{tr}(\Delta X \Delta S)$$

$$= (1 - \alpha)\operatorname{tr}(XS) + \alpha\sigma\mu n + \alpha\operatorname{tr}(R^r),$$

where the final equality follows from (5.26). Dividing the above expression by n and applying (5.23) yields (5.27).

---

[1]This is due to the fact that the eigenvalues of M are invariant under the transformation $H_P(M)$.

From here, we apply equations (5.22d), (5.22e), (5.27) and (5.22f), and it follows:

$$
\begin{aligned}
\widetilde{X}(\alpha)&\widetilde{S}(\alpha) - \mu(\alpha)I \\
&= (\widetilde{X} + \alpha\widetilde{\Delta X})(\widetilde{S} + \alpha\widetilde{\Delta S}) - \mu(\alpha)I \\
&= \widetilde{X}\widetilde{S} + \alpha(\widetilde{\Delta X}\widetilde{S} + \widetilde{X}\widetilde{\Delta S}) + \alpha^2\widetilde{\Delta X}\widetilde{\Delta S} - \mu(\alpha)I \\
&= (1-\alpha)\widetilde{X}\widetilde{S} + \alpha\widetilde{X}\widetilde{S} + \alpha(\widetilde{\Delta X}\widetilde{S} + \widetilde{X}\widetilde{\Delta S}) + \alpha^2\widetilde{\Delta X}\widetilde{\Delta S} - \mu(\alpha)I \\
&= (1-\alpha)\widetilde{X}\widetilde{S} + \alpha(\widetilde{\Delta X}\widetilde{S} + \widetilde{X}\widetilde{\Delta S} + \widetilde{X}\widetilde{S}) + \alpha^2\widetilde{\Delta X}\widetilde{\Delta S} - \mu(\alpha)I \\
&= (1-\alpha)\widetilde{X}\widetilde{S} + \alpha(\widetilde{\Delta X}\widetilde{S} + \widetilde{X}\widetilde{\Delta S} + \widetilde{X}\widetilde{S}) + \alpha^2\widetilde{\Delta X}\widetilde{\Delta S} - \mu(\alpha)I + \underbrace{(\alpha\widetilde{R}^r - \alpha\widetilde{R}^r)}_{=0} \\
&= (1-\alpha)\widetilde{X}\widetilde{S} + \alpha(\widetilde{\Delta X}\widetilde{S} + \widetilde{X}\widetilde{\Delta S} + \widetilde{X}\widetilde{S}) + \alpha^2\widetilde{\Delta X}\widetilde{\Delta S} - \left[(1-\alpha+\sigma\alpha)\mu + \frac{\alpha\operatorname{tr}(R^r)}{n}\right]I \\
&\quad + (\alpha\widetilde{R}^r - \alpha\widetilde{R}^r) \\
&= (1-\alpha)(\widetilde{X}\widetilde{S} - \mu I) + \alpha(\widetilde{\Delta X}\widetilde{S} + \widetilde{X}\widetilde{\Delta S} + \widetilde{X}\widetilde{S} - \sigma\mu I - \widetilde{R}^r) + \alpha^2\widetilde{\Delta X}\widetilde{\Delta S} \\
&\quad + \alpha\left[\widetilde{R}^r - \frac{\operatorname{tr}(R^r)}{n}I\right] \\
&= (1-\alpha)(\widetilde{X}\widetilde{S} - \mu I) + \alpha\widetilde{X}^{1/2}\mathcal{W}_X\widetilde{X}^{-1/2} + \alpha^2\widetilde{\Delta X}\widetilde{\Delta S} + \alpha\left[\widetilde{R}^r - \frac{\operatorname{tr}(R^r)}{n}I\right],
\end{aligned}
$$

which holds for any $\alpha \in \mathbb{R}$. Multiplying on the left by $\widetilde{X}^{-1/2}$ and on the right by $\widetilde{X}^{1/2}$ gives (5.28). $\qquad\square$

The following result from [Mon98] plays a crucial role in the analysis.

Lemma 5.9 (Lemma 3.5 in [Mon98]). *Let $M \in \mathbb{R}^{n\times n}$ be such that $H_Q(M) = 0$ for some nonsingular $Q \in \mathbb{R}^{n\times n}$. Then,*

$$\|H_I(M)\|_F \leq \frac{1}{2}\|M - M^\top\|_F, \tag{5.29}$$

$$\|M\|_F \leq \frac{\sqrt{2}}{2}\|M - M^\top\|_F. \tag{5.30}$$

In particular, if $M = U_1 + U_2$ for some $U_1 \in \mathcal{S}^n$ and $U_2 \in \mathbb{R}^{n\times n}$, then

$$\|M\|_F \leq \sqrt{2}\|U_2\|_F.$$

We can now use the above result in order to prove the following lemma, which is adapted from Lemma 3.6 in [Mon98].

Lemma 5.10. *For every $\theta \in \mathbb{R}$, we have*

$$\|\mathcal{W}_X\|_F \leq \sqrt{2}\delta_x\left\|\widetilde{S}^{1/2}\widetilde{X}^{1/2} - \theta\mu\widetilde{S}^{-1/2}\widetilde{X}^{-1/2}\right\| + \sqrt{2}\left\|\widetilde{X}^{-1/2}\widetilde{R}^r\widetilde{X}^{1/2}\right\|_F, \tag{5.31}$$

and

$$\left\|\widetilde{X}^{-1/2}\left[\widetilde{X}(\alpha)\widetilde{S}(\alpha) - \mu(\alpha)I\right]\widetilde{X}^{1/2}\right\|_F$$

$$\leq (1-\alpha)d(\widetilde{X},\widetilde{S}) + \alpha^2\delta_x\delta_s + \alpha\left\|\widetilde{X}^{-1/2}\left[\widetilde{R}^r - \frac{\operatorname{tr}(R^r)}{n}I\right]\widetilde{X}^{1/2}\right\|$$

$$+ \alpha\sqrt{2}\left\|\widetilde{S}^{1/2}\widetilde{X}^{1/2} - \theta\mu\widetilde{S}^{-1/2}\widetilde{X}^{-1/2}\right\| + \sqrt{2}\|\widetilde{X}^{-1/2}\widetilde{R}^r\widetilde{X}^{1/2}\|, \tag{5.32}$$

for all $\alpha \in [0,1]$, where

$$\delta_x = \left\|\widetilde{X}^{-1/2}\widetilde{\Delta X}\widetilde{S}^{1/2}\right\|_F, \quad \delta_s = \left\|\widetilde{S}^{-1/2}\widetilde{\Delta S}\widetilde{X}^{1/2}\right\|_F. \tag{5.33}$$

Proof. We follow a similar strategy to [Mon98], but modify the decomposition (namely the term U_2), to account for the inexactness residual term. Let $\theta \in \mathbb{R}$ be given. By (5.22f), we have $\mathcal{W}_X = U_1 + U_2$, where

$$U_1 = \widetilde{X}^{1/2}\widetilde{\Delta S}\widetilde{X}^{1/2} + \theta\mu\widetilde{X}^{-1/2}\widetilde{\Delta X}\widetilde{X}^{-1/2} + \widetilde{X}^{1/2}\widetilde{S}\widetilde{X}^{1/2} - \sigma\mu I \in \mathcal{S}^n, \text{ and}$$

$$U_2 = \widetilde{X}^{-1/2}\widetilde{\Delta X}\widetilde{S}^{1/2}\left(\widetilde{S}^{1/2}\widetilde{X}^{1/2} - \theta\mu\widetilde{S}^{-1/2}\widetilde{X}^{-1/2}\right) - \widetilde{X}^{-1/2}\widetilde{R}^r\widetilde{X}^{1/2}.$$

Then, applying (5.25), it follows $H_Q(\mathcal{W}_X) = 0$ for $Q = \widetilde{X}^{1/2}$. This fact, combined with Lemma 5.9 and the fact that $\|AB\|_F \leq \|A\|_F\|B\|$ for every $A, B \in \mathbb{R}^{n\times n}$, from the definition of U_2 and δ_x it follows

$$\|\mathcal{W}_X\|_F \leq \sqrt{2}\|U_2\|_F \leq \sqrt{2}\delta_x\left\|\widetilde{S}^{1/2}\widetilde{X}^{1/2} - \theta\mu\widetilde{S}^{-1/2}\widetilde{X}^{-1/2}\right\| + \sqrt{2}\left\|\widetilde{X}^{-1/2}\widetilde{R}^r\widetilde{X}^{1/2}\right\|_F,$$

i.e., (5.31) holds. From here, we apply (5.6) and (5.28), and note that for $\alpha \in [0,1]$, by the definition of δ_x and δ_s we have:

$$\left\|\widetilde{X}^{-1/2}\left[\widetilde{X}(\alpha)\widetilde{S}(\alpha) - \mu(\alpha)I\right]\widetilde{X}^{1/2}\right\|_F$$

$$\leq (1-\alpha)\left\|\widetilde{X}^{1/2}\widetilde{S}\widetilde{X}^{1/2} - \mu I\right\|_F + \alpha\|\mathcal{W}_X\|_F + \alpha\delta_x\delta_s + \alpha\left\|\widetilde{X}^{-1/2}\left[\widetilde{R}^r - \frac{\operatorname{tr}(R^r)}{n}I\right]\widetilde{X}^{1/2}\right\|$$

$$= (1-\alpha)d(\widetilde{X},\widetilde{S}) + \alpha\|\mathcal{W}_X\|_F + \alpha^2\delta_x\delta_s + \alpha\left\|\widetilde{X}^{-1/2}\left[\widetilde{R}^r - \frac{\operatorname{tr}(R^r)}{n}I\right]\widetilde{X}^{1/2}\right\|.$$

Combining the above result with (5.31) implies (5.32) and the proof is complete. $\qquad\square$

For ease of notation, we follow [Mon98] in defining the quantity

$$\Phi_\theta(A,B) = \left\|A^{1/2}B^{1/2} - \theta\frac{\operatorname{tr}(AB)}{n}A^{-1/2}B^{-1/2}\right\|_F \tag{5.34}$$

for every $(A,B) \in \mathcal{S}^n_{++} \times \mathcal{S}^n_{++}$ and $\theta \in \mathbb{R}$.

Lemma 5.11 (Lemma 3.7 in [Mon98]). *If $d(X, S) \leq \gamma\mu$ for some $\gamma \in (0, 1)$, then*

$$\left\| \tilde{X}^{-1/2} \tilde{S}^{-1/2} \right\|^2 \leq \frac{1}{(1 - \gamma)\mu}, \tag{5.35a}$$

$$\left[\Phi_\theta(\tilde{X}, \tilde{S}) \right]^2 \leq \frac{\gamma^2 + (1 - \theta)^2 n}{1 - \gamma} \mu. \tag{5.35b}$$

The following result establishes upper bounds on the Frobenius norms of terms involving the residuals.

Lemma 5.12. *Suppose that the error tolerances are chosen such that the assumptions* (AR1), (AR2) *and* (AR3) *hold. For $\beta \in (0, 1)$ we have*

$$\|R^r\|_F \leq \beta\gamma\sigma\mu, \quad \|\tilde{R}^r\|_F \leq \beta\gamma\sigma\mu, \tag{5.36a}$$

$$\left\| \tilde{X}^{-1/2} \tilde{R}^r \tilde{X}^{1/2} \right\|_F \leq \beta\gamma\sigma\mu, \tag{5.36b}$$

$$\left\| \tilde{X}^{-1/2} \left[\tilde{R}^r - \frac{\operatorname{tr}(R^r)}{n} I \right] \tilde{X}^{1/2} \right\|_F \leq 2\beta\gamma\sigma\mu. \tag{5.36c}$$

Proof. From equation (5.24c), we have $\tilde{R}^c = \sigma\mu I - H_I(\tilde{X}, \tilde{S})$. We first note that

$$\tilde{X}^{-1/2} \tilde{R}^c \tilde{X}^{1/2} = \tilde{X}^{-1/2} H_I(\sigma\mu I - \tilde{X}\tilde{S})\tilde{X}^{1/2} = \tilde{X}^{-1/2}[\sigma\mu I - \tilde{X}\tilde{S}]\tilde{X}^{1/2}$$

$$= \sigma\mu I - \tilde{X}^{-1/2}\tilde{X}\tilde{S}\tilde{X}^{1/2}$$

$$= \sigma\mu I - \tilde{X}^{1/2}\tilde{S}\tilde{X}^{1/2}.$$

Thus, it follows

$$\|\tilde{X}^{-1/2}\tilde{R}^c\tilde{X}^{1/2}\|_F = \|\sigma\mu I - \tilde{X}^{1/2}\tilde{S}\tilde{X}^{1/2}\|_F = \|\tilde{X}^{1/2}\tilde{S}\tilde{X}^{1/2} - \sigma\mu I\|_F = d_\sigma(\tilde{X}, \tilde{S}).$$

Further, applying Lemma 5.5(a) one can observe

$$\|\tilde{R}^c\|_F = \|H_I(\tilde{X}, \tilde{S}) - \sigma\mu I\|_F = d_\sigma(\tilde{X}, \tilde{S}) = d_\sigma(X, S) \leq \sigma\gamma\mu.$$

Combining this fact with assumptions (AR1) and (AR2) we have

$$\|R^r\|_F \leq \beta\|R^c\|_F \leq \beta\sigma\gamma\mu, \text{ and } \|\tilde{R}^r\|_F \leq \beta\|\tilde{R}^c\|_F \leq \beta\sigma\gamma\mu,$$

and hence (5.36a) holds. By assumption (AR3), it follows that

$$\|\tilde{X}^{-1/2}\tilde{R}^r\tilde{X}^{1/2}\|_F \leq \beta\|\tilde{X}^{-1/2}\tilde{R}^c\tilde{X}^{1/2}\|_F \leq \beta\sigma\gamma\mu.$$

Recall that for two matrices $A, B \in \mathbb{R}^{n \times n}$, $\operatorname{tr}(AB) \leq \|A\|_F \|B\|_F$, thus

$$\operatorname{tr}(R^r) = \operatorname{tr}(R^r I) \leq \|R^r\|_F \|I\|_F = \sqrt{n}\|R^r\|_F.$$

Then, noting that $\|R^r\|_F \leq \beta \|R^c\| \leq \beta\sigma\gamma\mu$, it follows

$$\left\| \tilde{X}^{-1/2}\left[\tilde{R}^r - \frac{\operatorname{tr}(R^r)}{n}I \right]\tilde{X}^{1/2}\right\|_F \leq \left\| \tilde{X}^{-1/2}\tilde{R}^r\tilde{X}^{1/2}\right\|_F + \left\| \frac{\operatorname{tr}(R^r)}{n}I\right\|_F$$

$$\leq \beta\sigma\gamma\mu + \frac{|\operatorname{tr}(R^r)|}{n}\sqrt{n}$$

$$\leq \beta\sigma\gamma\mu + \|R^r\|_F \leq 2\beta\sigma\gamma\mu,$$

which completes the proof. $\qquad\square$

In the next result, we establish an upper bound on the terms δ_x and δ_s by adapting Lemma 3.8 of [Mon98] to the inexact setting.

Lemma 5.13. *If $(X, y, S) \in \mathcal{N}_F(\gamma)$ for some $\gamma > 0$ satisfying*

$$2\sqrt{2}\frac{\gamma}{1-\gamma} \leq 1, \tag{5.37}$$

and $\beta \in (0,1)$ such that

$$\beta\sigma \leq \sqrt{\frac{\gamma^2 + (1-\sigma)^2 n}{1-\gamma}},$$

then

$$\max\{\delta_x, \delta_s\} \leq 2\left(\Phi_\sigma(\tilde{X}, \tilde{S}) + \sqrt{\frac{\gamma^2 + (1-\sigma)^2 n}{1-\gamma}}\mu \right),$$

where δ_x, δ_s and $\Phi_\sigma(\cdot, \cdot)$ are defined by (5.33) and (5.34).

Proof. From (5.22f), one can see

$$\tilde{X}^{1/2}\widehat{\Delta S}\tilde{S}^{-1/2} + \tilde{X}^{-1/2}\widehat{\Delta X}\tilde{S}^{1/2} = W_X\tilde{X}^{-1/2}\tilde{S}^{-1/2} + \sigma\mu\tilde{X}^{-1/2}\tilde{S}^{-1/2} - \tilde{X}^{1/2}\tilde{S}^{1/2}$$

$$+ (\tilde{X}^{-1/2}\tilde{R}^r\tilde{X}^{1/2})\tilde{X}^{-1/2}\tilde{S}^{-1/2}. \tag{5.38}$$

Since $\gamma \in (0,1)$, it follows that $(1-\gamma) \in (0,1)$ so we have $(1-\gamma) < (1-\gamma)^{1/2}$ and hence

$$\frac{1}{(1-\gamma)^{1/2}} < \frac{1}{1-\gamma}.$$

From (5.26), observe that the two terms on the left hand side of (5.38) are orthogonal. Combining this fact with (5.33), (5.38), (5.34), (5.23), (5.31) with $\theta = 1$; and (5.35a), (5.35b) with $\theta = 1$ and (5.37), it follows that:

$$\max\{\delta_x, \delta_s\} \leq \left(\delta_x^2 + \delta_s^2 \right)^{1/2} = \left(\left\| \tilde{X}^{-1/2}\widehat{\Delta X}\tilde{S}^{1/2}\right\|_F^2 + \left\| \tilde{S}^{-1/2}\widehat{\Delta S}\tilde{X}^{1/2}\right\|_F^2 \right)^{1/2}$$

$$= \left\| \tilde{X}^{-1/2}\widehat{\Delta X}\tilde{S}^{1/2} + \tilde{X}^{1/2}\widehat{\Delta S}\tilde{S}^{-1/2}\right\|_F$$

$$\leq \|W_X\|_F \left\| \tilde{X}^{-1/2}\tilde{S}^{-1/2}\right\| + \left\| \tilde{X}^{1/2}\tilde{S}^{1/2} - \sigma\mu\tilde{X}^{-1/2}\tilde{S}^{-1/2}\right\|$$

$$+ \left\| (\tilde{X}^{-1/2}\tilde{R}^r\tilde{X}^{1/2})\tilde{X}^{-1/2}\tilde{S}^{-1/2}\right\|_F \quad \text{(by (5.38))}$$

$$= \|W_X\|_F \left\|\widetilde{X}^{-1/2}\widetilde{S}^{-1/2}\right\| + \left\|\widetilde{X}^{1/2}\widetilde{S}^{1/2} - \sigma\mu\widetilde{X}^{-1/2}\widetilde{S}^{-1/2}\right\| + \left\|\widetilde{X}^{-1/2}\widetilde{R}^r\widetilde{S}^{-1/2}\right\|_F$$

$$\leq \sqrt{2}\delta_x \left\|\widetilde{X}^{1/2}\widetilde{S}^{1/2} - \mu\widetilde{X}^{-1/2}\widetilde{S}^{-1/2}\right\|_F \left\|\widetilde{X}^{-1/2}\widetilde{S}^{-1/2}\right\| + \Phi_\sigma(\widetilde{X},\widetilde{S})$$

$$+ (1+\sqrt{2}) \left\|\widetilde{X}^{-1/2}\widetilde{R}^r\widetilde{X}^{1/2}\right\|_F \left\|\widetilde{X}^{-1/2}\widetilde{S}^{-1/2}\right\|$$

$$\leq \sqrt{2}\delta_x \frac{\gamma\mu^{1/2}}{(1-\gamma)^{1/2}} \frac{1}{(1-\gamma)^{1/2}\mu^{1/2}} + \Phi_\sigma(\widetilde{X},\widetilde{S}) + (1+\sqrt{2})\beta\sigma\gamma\mu\frac{1}{(1-\gamma)^{1/2}\mu^{1/2}}$$

$$= \sqrt{2}\delta_x \frac{\gamma\mu^{1/2}}{(1-\gamma)^{1/2}} \frac{1}{(1-\gamma)^{1/2}\mu^{1/2}} + \Phi_\sigma(\widetilde{X},\widetilde{S}) + \left[(1+\sqrt{2})\frac{\gamma}{(1-\gamma)^{1/2}}\right]\beta\sigma\mu^{1/2}$$

$$< \sqrt{2}\delta_x \frac{\gamma\mu^{1/2}}{(1-\gamma)^{1/2}} \frac{1}{(1-\gamma)^{1/2}\mu^{1/2}} + \Phi_\sigma(\widetilde{X},\widetilde{S}) + \left[(1+\sqrt{2})\frac{\gamma}{(1-\gamma)}\right]\beta\sigma\mu^{1/2}$$

$$\leq \sqrt{2}\delta_x \frac{\gamma\mu^{1/2}}{(1-\gamma)^{1/2}} \frac{1}{(1-\gamma)^{1/2}\mu^{1/2}} + \Phi_\sigma(\widetilde{X},\widetilde{S}) + \left[2\sqrt{2}\frac{\gamma}{(1-\gamma)}\right]\beta\sigma\mu^{1/2}$$

$$\leq \sqrt{2}\delta_x \frac{\gamma\mu^{1/2}}{(1-\gamma)^{1/2}} \frac{1}{(1-\gamma)^{1/2}\mu^{1/2}} + \Phi_\sigma(\widetilde{X},\widetilde{S}) + \beta\sigma\mu^{1/2}$$

$$\leq \sqrt{2}\delta_x \frac{\gamma\mu^{1/2}}{(1-\gamma)^{1/2}} \frac{1}{(1-\gamma)^{1/2}\mu^{1/2}} + \Phi_\sigma(\widetilde{X},\widetilde{S}) + \sqrt{\frac{\gamma^2 + (1-\sigma)^2 n}{1-\gamma}}\mu$$

$$\leq \frac{\delta_x}{2} + \Phi_\sigma(\widetilde{X},\widetilde{S}) + \sqrt{\frac{\gamma^2 + (1-\sigma)^2 n}{1-\gamma}}\mu$$

$$\leq \frac{1}{2}\max\{\delta_x,\delta_s\} + \Phi_\sigma(\widetilde{X},\widetilde{S}) + \sqrt{\frac{\gamma^2 + (1-\sigma)^2 n}{1-\gamma}}\mu,$$

and the proof is complete. $\square$

The final result of this section, analogous to Lemma 3.9 of [Mon98], follows from the above lemmas.

Lemma 5.14. *If* $(X,y,S) \in \mathcal{N}_F(\gamma)$ *for some* $\gamma > 0$ *satisfying* (5.37), *then, for every* $\alpha \in [0,1]$, $\beta \in (0,1)$, *and* $\eta > 0$, *we have*

$$\left\|\widetilde{X}^{-1/2}\left[\widetilde{X}(\alpha)\widetilde{S}(\alpha) - \mu(\alpha)I\right]\widetilde{X}^{1/2}\right\|_F$$

$$\leq \left((1-\alpha)\gamma + 4\sqrt{2}\alpha\frac{\gamma\left[\gamma^2 + (1-\sigma)^2 n\right]^{1/2}}{1-\gamma} + 16\alpha^2\frac{\gamma^2 + (1-\sigma)^2 n}{1-\gamma} + \alpha(2+\sqrt{2})\beta\sigma\gamma\right)\mu.$$

Proof. We apply (5.32) with $\theta = 1$, Lemma 5.5 (a), (5.22a), the assumption $d(X,S) \leq \gamma\mu$,

Lemma 5.12, Lemma 5.13 and (5.35b) where we choose $\theta = 1$ and $\theta = \sigma$. We have:

$$\left\| \widetilde{X}^{-1/2} \left[\widetilde{X}(\alpha)\widetilde{S}(\alpha) - \mu(\alpha)I \right] \widetilde{X}^{1/2} \right\|_F$$

$$\leq (1-\alpha)d(X,S) + \sqrt{2}\alpha\delta_x\Phi_1(\widetilde{X},\widetilde{S}) + \alpha^2\delta_x\delta_s + \alpha(2+\sqrt{2})\beta d_\sigma(X,S)$$

$$\leq (1-\alpha)\gamma\mu + 2\sqrt{2}\alpha \left(\Phi_\sigma(\widetilde{X},\widetilde{S}) + \sqrt{\frac{\gamma^2 + (1-\sigma)^2 n}{1-\gamma}}\mu \right) \Phi_1(\widetilde{X},\widetilde{S})$$

$$+ \alpha^2 \left[2\Phi_\sigma(\widetilde{X},\widetilde{S}) + 2\sqrt{\frac{\gamma^2 + (1-\sigma)^2 n}{1-\gamma}}\mu \right]^2 + \alpha(2+\sqrt{2})\beta\sigma\gamma\mu$$

$$\leq (1-\alpha)\gamma\mu + 2\sqrt{2}\alpha \left(2\sqrt{\frac{\gamma^2 + (1-\sigma)^2 n}{1-\gamma}}\mu \right) \Phi_1(\widetilde{X},\widetilde{S}) + \alpha^2 \left[4\sqrt{\frac{\gamma^2 + (1-\sigma)^2 n}{1-\gamma}}\mu \right]^2 + \alpha(2+\sqrt{2})\beta\sigma\gamma\mu$$

$$= (1-\alpha)\gamma\mu + 4\sqrt{2}\alpha\sqrt{\frac{\gamma^2 + (1-\sigma)^2 n}{1-\gamma}}\mu \cdot \Phi_1(\widetilde{X},\widetilde{S}) + 16\alpha^2 \left[\sqrt{\frac{\gamma^2 + (1-\sigma)^2 n}{1-\gamma}}\mu \right]^2 + \alpha(2+\sqrt{2})\beta\sigma\gamma\mu$$

$$\leq \left((1-\alpha)\gamma + 4\sqrt{2}\alpha\frac{\gamma\left[\gamma^2 + (1-\sigma)^2 n\right]^{1/2}}{1-\gamma} + 16\alpha^2\frac{\gamma^2 + (1-\sigma)^2 n}{1-\gamma} + \alpha(2+\sqrt{2})\beta\sigma\gamma \right) \mu.$$

The proof is complete. $\qquad\square$

5.5 Quantum Interior Point Methods for SDO

In this section we present the Inexact-Feasible QIPM. For the IF-QIPM, we establish polynomial convergence, then discuss how to implement the block-encoding of the Newton linear system. The overall running time analysis is conducted in Section 5.6.

5.5.1 An Inexact-Feasible QIPM for SDO

We quantize a general IF-IPM and the resulting scheme is described in Algorithm 4. There is no classical counterpart in the literature for SDO, as the nullspace representation of the Newton system destroys the symmetry of the Newton system, and is thus impractical for classical computing. The algorithm has several parameters, the most important of which is the optimality gap tolerance ϵ. Details for all steps of Algorithm 4 are discussed subsequently in this paper.

In our initialization steps, we must calculate bases of the nullspace and rowspace of $\mathcal{A}_s$. For this step, one can either use Gaussian elimination, or the QR-factorization of $\mathcal{A}_s$. This can can be accomplished using $\mathcal{O}(n^{2\omega})$ classical arithmetic operations where $\omega = 2.37$ is the matrix multiplication exponent [BCS13, Str69], although this cost can be potentially reduced much closer to $\mathcal{O}(n^4)$ if the initial data is sparse and we use sparse matrix-vector multiplication, see [YZ05]. We also point out that the QR-factorization has been shown to exhibit a higher level of numerical stability compared to Gaussian elimination. It is also possible to trivially obtain a basis for the nullspace and the range space for "free" if the SDO problem is given in canonical form (inequalities instead of equalities), in which case, one could use the Self-Dual embedding model for that form, the nullspace basis matrix can be constructed trivially too from the coefficient matrix (without needing a QR-factorization or Gaussian elimination). Crucially, the computation of these bases need only be carried out once. That is, we calculate $\mathcal{A}_s$ and Q_2 one time, and store these matrices in QRAM before the algorithm begins, as they remain unchanged for the duration of the algorithm. This step is akin to the preprocessing steps (e.g., Cholesky factorization, ordering of basis) typical of many optimization algorithms.

In steps 2 and 3 of Algorithm 4 we require classical matrix multiplication and inversion of $n \times n$ matrices. These steps require $\mathcal{O}(n^\omega)$ classical arithmetic operations. A detailed analysis for step 3 can be found in Proposition 5.20. In step 4, we employ the Iterative Refinement scheme for linear systems studied in Chapter 4 routine which repeatedly applies a QLSA and quantum state tomography to solve a sequence of systems whose solutions can be used to construct a classical description of the solution to the Newton linear system. Following our results on IR for the LSP in Section 4.4.1, this can be accomplished in time $\mathcal{O}\left(\left(n^2\alpha\kappa\varrho + n^4\right) \cdot \mathrm{polylog}\left(n, \kappa, \frac{1}{\xi^{(k)}}\right)\right)$, when the data is stored in QRAM, where $\xi^{(k)}$ is the ℓ_2-norm error to which we extract a classical estimate of the solution to the Newton linear system, ϱ is a norm bound on the solution, and $\alpha = \|M_{NT}\|_F$ is the subnormalization factor used to block-encode the Newton system coefficient matrix. Note that $\xi^{(k)}$ is chosen according to a sequence described in Section 5.5.1.1. Then, in step 5, we use our classical estimates of $\overline{\Delta z}$ and $\overline{\Delta y}$ to classically calculate $\overline{\Delta X}$ and $\overline{\Delta S}$. This step amounts to matrix-vector multiplication of dimension n^2, and requires $\mathcal{O}(n^4)$ arithmetic operations. Finally, we classically update the current solutions to the SDO problems $X^{(k)}$, $y^{(k)}$ and $S^{(k)}$, and the central path parameter $\mu^{(k)}$.

Here we establish that Algorithm 4 converges for a specific choice of the error parameters, discuss how to construct the Newton linear system for the Nesterov-Todd direction, and provide a complexity analysis for the chosen parameters.

5.5.1.1 Polynomial Convergence of the IF-QIPM

In order to have a convergent algorithm, we need to ensure that the errors ξ_k introduced by tomography are properly controlled. Now, by definition $\|R^c\|_F \leq \sigma\gamma\mu$ holds at every iteration, and we terminate once $\mu \leq \epsilon$. Hence, for every iteration of the algorithm before termination with an ϵ-optimal solution, the inequality

$$\|R^c\|_F > \sigma\gamma\epsilon,$$

holds, where $\gamma = \frac{1}{20}$ and $\sigma \approx 1$ for large n. In other words, the requirement on the residual bound never becomes much smaller than ϵ, if we aim at ϵ-optimality. Thus, we can set

$$\xi^{(k)} = \frac{\beta}{\varrho} \cdot \max\left\{ \|R^c\|_F, \frac{1}{25}\epsilon\right\}. \tag{5.39}$$

Algorithm 4 Inexact-Feasible Quantum Interior Point Method

Input: $\epsilon, \delta > 0$; $\sigma = 1 - \delta/\sqrt{n}$; $\beta, \gamma \in (0, 1)$

Choose a nonsingular matrix P from Monteiro-Zhang family of search directions

Choose $(X^{(0)}, y^{(0)}, S^{(0)}) \in \mathcal{N}_F(\gamma)$

Set $\mu^{(0)} = \dfrac{\mathrm{tr}\left(X^{(0)} S^{(0)}\right)}{n}$

Store $(X^{(0)}, y^{(0)}, S^{(0)})$ in QRAM

Compute bases for $\mathcal{R}(\mathcal{A}_s) = \mathcal{A}_s^\top$ and $\mathrm{Null}(\mathcal{A}_s) = Q_2$

Store $\mathcal{A}_s$ and Q_2 in QRAM

while $\mu > \epsilon$ **do**

    1. $\mu^{(k)} \leftarrow \dfrac{\mathrm{tr}\left(X^{(k)} S^{(k)}\right)}{n}$

    2. Compute matrices $(X^{(k)})^{-1}$ and $(S^{(k)})^{-1}$ classically. Store these matrices in QRAM.

    3. Compute scaling matrix $P^{(k)}$, right hand side matrix, and necessary matrix products for Newton system (5.19) and store in QRAM

    4. Solve (5.19) and obtain classical estimate $\overline{\Delta z}^{(k)}, \overline{\Delta y}^{(k)}$ of $\Delta z^{(k)}, \Delta y^{(k)}$ using Algorithm 2

    5. Use classical estimate $\overline{\Delta z}^{(k)}, \overline{\Delta y}^{(k)}$ to obtain classical estimate $\overline{\Delta X}^{(k)}, \overline{\Delta S}^{(k)}$ of $\Delta X^{(k)}, \Delta S^{(k)}$

    6. Update current solution and central path parameter

$$X^{(k+1)} \leftarrow X^{(k)} + \Delta X^{(k)}, \ S^{(k+1)} \leftarrow S^{(k)} + \Delta S^{(k)} \text{ and } y^{(k+1)} \leftarrow y^{(k)} + \Delta y^{(k)}$$

$$k \leftarrow k + 1$$

end

---

where ϱ is the maximum norm of a solution to the Newton linear system. The next result provides an upper bound for ϱ.

Proposition 5.15. *The norm ϱ of the solution of the quantum Newton linear system* (5.19) *satisfies*

$$\varrho = \mathcal{O}\left(\|M_{NT}^{-1}\|\right).$$

Proof. We have

$$\begin{bmatrix} \Delta z \\ \Delta y \end{bmatrix} = (M_{NT})^{-1} \mathrm{svec}(R^c + R^r).$$

By Lemma 5.12, we have

$$\|R^c + R^r\|_F \leq \|R^c\|_F + \|R^r\|_F \leq (1 + \beta)\sigma\gamma\mu. \tag{5.40}$$

Choosing a constant $c_0 \geq (1 + \beta)\sigma\gamma$, it follows that $\|R^c + R^r\|_F \leq c_0\nu$, and therefore

$$\|R^c + R^r\|_F = \mathcal{O}(\mu) = \mathcal{O}(1),$$

where the final equality follows from the definition of condition number and $\mu = \mathcal{O}(1)$. To see this, note that across every iteration of the QIPM, we have $\max_k\{\mu^{(k)}\} = \mu^{(0)} = \dfrac{\mathrm{tr}\left(X^{(0)} S^{(0)}\right)}{n} =$

$\mathcal{O}(1)$, as in each iteration we strictly decrease the quantity $\mathrm{tr}\left(X^{(k)}S^{(k)}\right)$ (which consequently decreases μ).

The bound on the norm of the solution is thus given by

$$\|(\Delta z, \Delta y)\|_F = \|M_{NT}^{-1}\,\mathrm{svec}(R^c + R^r)\|_F \leq \|M_{NT}^{-1}\| \cdot \|(R^c + R^r)\|_F = \mathcal{O}\left(\|M_{NT}^{-1}\|\right).$$

The proof is complete. $\qquad\square$

Lemma 5.16. *Choosing $\xi^{(k)}$ according to (5.39), the inexact search direction computed at step 4 of Algorithm 4 satisfies* (AR1).

Proof. For feasible IPMs, the complementarity gap is updated as $\mu_{k+1} = (1 - \frac{\delta}{\sqrt{n}})\mu_k$, which implies that $\|R^c_{k+1}\|_F \leq (1 - \frac{\delta}{\sqrt{n}})\|R^c_k\|_F$, due to the fact that $\|R^c_{k+1}\|_F \leq \sigma\gamma\mu_{k+1}$. $\qquad\square$

Next, we provide the inexact analogue of Theorem 4.1 from [Mon98], which yields the analysis of one iteration of Algorithm 4 for suitable choices of γ, δ and β.

Theorem 5.17. *Let $\gamma, \beta \in (0,1)$ and $\delta \in [0, n^{1/2})$ be constants satisfying*

$$\frac{2\sqrt{2}\gamma}{1-\gamma} \leq 1, \quad \beta\sigma \leq \sqrt{\frac{\gamma^2 + (1-\sigma)^2 n}{1-\gamma}}, \quad \beta \leq 1 - \frac{\gamma}{\sqrt{n}} - \frac{21.7(\gamma^2 + \delta^2)}{(2+\sqrt{2})\left(1 - \frac{\delta}{\sqrt{n}}\right)\gamma(1-\gamma)}. \tag{5.41}$$

Suppose that $(X, y, S) \in \mathcal{N}_F(\gamma)$ and let $(\Delta X, \Delta y, \Delta S)$ denote the solution that we obtain from solving system (5.19), where $\sigma = 1 - \delta/\sqrt{n}$, $\mu = \mathrm{tr}(XS)/n$ and $P \in \mathbb{R}^{n \times n}$ is a nonsingular matrix. Then,

(a) $(\hat{X}, \hat{y}, \hat{S}) = (X + \Delta X, y + \Delta y, S + \Delta S) \in \mathcal{N}_F(\gamma)$;

(b) $\mathrm{tr}\left(\hat{X}\hat{S}\right) = \left(1 - \frac{\delta}{\sqrt{n}}\right)\mathrm{tr}(XS)$.

Proof. In the following proof, we seek to show that a new iterate, after a step α in this direction remains in the neighborhood of the central path. That is, the new iterate satisfies

$$\left\|\widetilde{X}^{-1/2}\left[\widetilde{X}(\alpha)\widetilde{S}(\alpha) - \mu(\alpha)I\right]\widetilde{X}^{1/2}\right\|_F \leq \gamma\mu(\alpha) = \gamma(1 - \alpha + \sigma\alpha)\mu + \alpha\gamma\frac{\mathrm{tr}(R^r)}{n}.$$

Now, $|\mathrm{tr}(R^r)| \leq \sqrt{n}\|R^r\|_F \leq \sqrt{n}\beta\sigma\gamma\mu < \sqrt{n}\sigma\gamma\mu$. However, since we cannot make any conclusions regarding the sign of the quantity $\mathrm{tr}(R^r)$, it suffices to show

$$\left\|\widetilde{X}^{-1/2}\left[\widetilde{X}(\alpha)\widetilde{S}(\alpha) - \mu(\alpha)I\right]\widetilde{X}^{1/2}\right\|_F \leq \gamma(1 - \alpha + \sigma\alpha)\mu - \alpha\gamma\frac{\sigma\gamma\mu}{\sqrt{n}}. \tag{$*$}$$

First, we establish an upper bound on the quantity

$$\left\|\widetilde{X}^{-1/2}\left[\widetilde{X}(\alpha)\widetilde{S}(\alpha) - \mu(\alpha)I\right]\widetilde{X}^{1/2}\right\|_F,$$

and show that this upper bound satisfies the inequality $(*)$ for appropriately chosen γ, β and δ. Applying Lemma 5.14, along with the definition of σ and (5.41), for all $\alpha \in [0,1]$ we have

$$
\left\| \widetilde{X}^{-1/2} \left[\widetilde{X}(\alpha)\widetilde{S}(\alpha) - \mu(\alpha)I \right] \widetilde{X}^{1/2} \right\|_F
$$

$$
\leq \left((1-\alpha)\gamma + 4\sqrt{2}\alpha \frac{\gamma \left[\gamma^2 + (1-\sigma)^2 n \right]^{1/2}}{1-\gamma} + 16\alpha^2 \frac{\gamma^2 + (1-\sigma)^2 n}{1-\gamma} + \alpha(2+\sqrt{2})\beta\sigma\gamma \right) \mu
$$

$$
\leq \left((1-\alpha)\gamma + 21.7\alpha \frac{\gamma^2 + (1-\sigma)^2 n}{1-\gamma} + \alpha(2+\sqrt{2})\beta\sigma\gamma \right) \mu
$$

$$
= \left((1-\alpha)\gamma + \alpha \frac{21.7(\gamma^2 + (1-\sigma)^2 n)}{1-\gamma} + \alpha(2+\sqrt{2})\beta\sigma\gamma \right) \mu
$$

$$
= \left((1-\alpha)\gamma + \alpha \frac{21.7(\gamma^2 + \delta^2)}{1-\gamma} + \alpha(2+\sqrt{2})\beta\sigma\gamma \right) \mu. \quad (**)
$$

Substituting the upper bound $(**)$ into the inequality $(*)$, one can see that we must have

$$
\left(\alpha \frac{21.7(\gamma^2 + \delta^2)}{1-\gamma} + \alpha(2+\sqrt{2})\beta\sigma\gamma \right) \mu \leq \alpha \left(1 - \frac{\delta}{\sqrt{n}} \right) \gamma\mu - \alpha\gamma \frac{\sigma\gamma\mu}{\sqrt{n}}.
$$

Dropping like terms on both sides of the inequality, this implies

$$
\frac{21.7(\gamma^2 + \delta^2)}{1-\gamma} + (2+\sqrt{2})\beta\sigma\gamma \leq \left(1 - \frac{\delta}{\sqrt{n}} \right) \gamma - \frac{\sigma\gamma^2}{\sqrt{n}}.
$$

Using the fact that $\sigma = 1 - \delta/\sqrt{n}$, we can further simplify this expression to

$$
\beta \leq 1 - \frac{\gamma}{\sqrt{n}} - \frac{21.7(\gamma^2 + \delta^2)}{(2+\sqrt{2})\left(1 - \frac{\delta}{\sqrt{n}} \right)\gamma(1-\gamma)}. \tag{5.42}
$$

Choosing $\beta \in (0,1)$ such that (5.42) is satisfied, it follows

$$
\left\| \widetilde{X}^{-1/2} \left[\widetilde{X}(\alpha)\widetilde{S}(\alpha) - \mu(\alpha)I \right] \widetilde{X}^{1/2} \right\|_F \leq \left((1-\alpha)\gamma + \alpha \left(1 - \frac{\delta}{\sqrt{n}} \right) \gamma \right) \mu + \alpha\gamma \frac{\operatorname{tr}(R^r)}{n}
$$

$$
= \left((1-\alpha)\gamma + \alpha\gamma\sigma \right) \mu + \alpha\gamma \frac{\operatorname{tr}(R^r)}{n}. \tag{5.43}
$$

By equation (5.27)

$$
\mu(\alpha) = (1 - \alpha + \sigma\alpha)\mu + \alpha \frac{\operatorname{tr}(R^r)}{n}.
$$

Hence,

$$
\left\| \widetilde{X}^{-1/2} \left[\widetilde{X}(\alpha)\widetilde{S}(\alpha) - \mu(\alpha)I \right] \widetilde{X}^{1/2} \right\|_F \leq \gamma(1 - \alpha + \alpha\sigma)\mu + \gamma \frac{\operatorname{tr}(R^r)}{n} = \gamma\mu(\alpha).
$$

Therefore, defining

$$
G(\alpha) = \left(\widetilde{X}^{-1/2} \widetilde{X}(\alpha)\widetilde{S}(\alpha)\widetilde{X}^{1/2} \right) / \mu(\alpha),
$$

it follows that $G(\alpha)$ is nonsingular for all $\alpha \in [0,1]$ as $\|G(\alpha) - I\| \le \gamma < 1$. As a consequence, $\widetilde{X}(\alpha)$ and $\widetilde{S}(\alpha)$ are also nonsingular for $\alpha \in [0,1]$.

From here, we follow [Mon98] by combining (5.22d) and (5.22e) with a simple continuity argument which demonstrates that for each $\alpha \in [0,1]$, we have that $(\widetilde{X}(\alpha), \widetilde{S}(\alpha))$ and $(X(\alpha), S(\alpha))$ are in $\mathcal{S}^n_{++} \times \mathcal{S}^n_{++}$. Noting our assumption $(X, y, S) \in \mathcal{P}^0 \times \mathcal{D}^0$, along with equations (5.20b) and (5.20a), one can observe that $(X(\alpha), y(\alpha), S(\alpha))$ satisfies

$$\mathrm{tr}\,(A_i X) - b_i = 0, \quad i = 1, \ldots, m,$$

$$\sum_{i \in [m]} y_i A_i + S - C = 0,$$

and thus $(X(\alpha), y(\alpha), S(\alpha)) \in \mathcal{P}^0 \times \mathcal{D}^0$ for every $\alpha \in [0,1]$. Finally, we apply Lemma 5.5(a) with $(X, S) = (X(1), S(1))$ and $Q = P$, along with Lemma 5.5(b) with $(X, S) = (\widetilde{X}(1), \widetilde{S}(1))$ and $Q = \widetilde{X}^{-1/2}$, and equations (5.22d), (5.22e), (5.27) with $\alpha = 1$, and (5.43) with $\alpha = 1$, which yields

$$
\begin{aligned}
d\left(X(1), S(1)\right) = d\left(\widetilde{X}(1), \widetilde{S}(1)\right) \\
\le \left\| H_{\widetilde{X}^{-1/2}}\left[\widetilde{X}(1)\widetilde{S}(1) - \mu(1)I\right] \right\|_F \\
\le \left\| \widetilde{X}^{-1/2}\left[\widetilde{X}(1)\widetilde{S}(1) - \mu(1)I\right]\widetilde{X}^{1/2} \right\|_F \le \gamma\mu(1).
\end{aligned}
$$

Therefore, $(\widehat{X}, \hat{y}, \widehat{S}) = (X + \Delta X, y + \Delta y, S + \Delta S) \in \mathcal{N}_F(\gamma)$, and hence (a) holds. Further, (b) follows immediately from the definition of σ and equation (5.27) with $\alpha = 1$. $\qquad \square$

Following from Theorem 5.17, we obtain the convergence result for Algorithm 4.

Corollary 5.1. *Let γ, δ and β be constants in $(0,1)$ and satisfying*

$$\frac{2\sqrt{2}\gamma}{1-\gamma} \le 1, \quad \beta\sigma \le \sqrt{\frac{\gamma^2 + (1-\sigma)^2 n}{1-\gamma}}, \quad \beta \le 1 - \frac{\gamma}{\sqrt{n}} - \frac{21.7(\gamma^2 + \delta^2)}{(2 + \sqrt{2})\left(1 - \frac{\delta}{\sqrt{n}}\right)\gamma(1-\gamma)}.$$

Then, every iterate $\left(X^{(k)}, y^{(k)}, S^{(k)}\right)$ generated by the IF-QIPM in Algorithm 4 is an element of the neighborhood $\mathcal{N}_F(\gamma)$, and satisfies the relation $\mathrm{tr}\left(X^{(k)} S^{(k)}\right) = (1 - \delta/\sqrt{n})^k \left[\mathrm{tr}\left(X^{(0)} S^{(0)}\right)\right]$. Further, the IF-QIPM terminates in at most $\mathcal{O}(\sqrt{n}\log(1/\epsilon))$ iterations.

Examples of constants γ, δ and β satisfying the conditions of Corollary 5.1 are $\gamma = \delta = 1/20$ and $\beta = 1/4$.

5.5.1.2 Implementing the Nesterov-Todd linear system

As in [KP18], we factorize the NT Newton linear system given by (5.19), and construct block-encodings for its factors. In our algorithm, we use the Iterative Refinmenet method for linear systems described in Chapter 4 within an IPM, and to fully leverage the possible speedups we pre-compute some matrix inverses and products *classically*, and store the results in QRAM. This is more efficient than other schemes that we explored. Note that the largest matrices appearing in the Newton system, namely $Q_2 \in \mathbb{R}^{\frac{n(n+1)}{2} \times (\frac{n(n+1)}{2} - m)}$ and $\mathcal{A}_s^\top \in \mathbb{R}^{\frac{n(n+1)}{2} \times m}$, are only calculated once before the algorithm begins so their size is not prohibitive. Hence, compared to solving the Newton linear system (natrually an $n^2 \times n^2$ matrix), classical computation of matrix powers and products of $n \times n$ matrices has an insignificant cost, and performing these steps classically allows also us to circumvent increased dependence on the condition number.

Proposition 5.18. *Let*

$$M_1 = \begin{bmatrix} (P \otimes_s P^{-\top} S)Q_2 & 0 \end{bmatrix}.$$

Let P, Q_2, and $P^{-\top} S$ be stored in QRAM. Then a $(\|M_1\|_F, \mathcal{O}(\log n), \xi_{M_1})$-block-encoding of M_1 can be constructed in time $\widetilde{\mathcal{O}}_{\frac{n}{\xi_{M_1}}}(1)$.

Proof. First, given that Q_2 is stored in QRAM, we can construct a $(\|Q_2\|_F, \mathcal{O}(\log n), \xi_1)$-block-encoding of Q_2 in time $\widetilde{\mathcal{O}}_{\frac{n}{\xi_1}}(1)$. Next, with P and $P^{-\top} S$ stored in QRAM, we can apply Proposition 3.10 so that we can construct a $(\|P \otimes_s P^{-\top} S\|_F, \mathcal{O}(\log n), \xi_2)$-block-encoding of $P \otimes_s P^{-\top} S$ in time $\widetilde{\mathcal{O}}_{\frac{n}{\xi_2}}(1)$.

From here we apply Proposition 3.6 with

$$\xi_1 = \frac{\xi_{M_1}}{2\|P \otimes_s P^{-\top} S\|_F} \text{ and } \xi_2 = \frac{\xi_{M_1}}{2\|Q_2\|_F},$$

yielding a $(\|M_1\|_F, \mathcal{O}(\log n), \xi_{M_1})$-block-encoding of M_1 in time $\widetilde{\mathcal{O}}_{\frac{n}{\xi_{M_1}}}(1)$, as desired. $\square$

Proposition 5.19. *Let*

$$M_2 = \begin{bmatrix} 0 & (PX \otimes_s P^{-\top})(-\mathcal{A}_s^\top) \end{bmatrix}.$$

Let $P^{-\top}$, $\mathcal{A}_s$, PX and $P^{-\top} S$ be stored in QRAM. Then a $(\|M_2\|_F, \mathcal{O}(\log n), \xi_{M_2})$-block-encoding of M_2 can be constructed in time $\widetilde{\mathcal{O}}_{\frac{n}{\xi_{M_2}}}(1)$.

Proof. The steps of the proof follow that of the previous result. With $-\mathcal{A}_s^\top$ stored in QRAM, we can construct a $(\|\mathcal{A}_s^\top\|_F, \mathcal{O}(\log n), \xi_1)$-block-encoding of $-\mathcal{A}_s^\top$ in time $\widetilde{\mathcal{O}}_{\frac{n}{\xi_1}}(1)$. Further, with PX and $P^{-\top}$ stored in QRAM, we apply Proposition 3.10 such that we can construct a $(\|PX \otimes_s P^{-\top}\|_F, \mathcal{O}(\log n), \xi_2)$-block-encoding of $PX \otimes_s P^{-\top}$ in time $\widetilde{\mathcal{O}}_{\frac{n}{\xi_2}}(1)$.

Applying Proposition 3.6 with

$$\xi_1 = \frac{\xi_{M_1}}{2\|PX \otimes_s P^{-\top}\|_F} \text{ and } \xi_2 = \frac{\xi_{M_1}}{2\|\mathcal{A}\|_F},$$

we obtain a $(\|M_2\|_F, \mathcal{O}(\log n), \xi_{M_2})$-block-encoding of M_2, and the proof is complete. $\square$

Proposition 5.20. *The Nesterov-Todd linear system matrix (5.19) can be written compactly as:*

$$M_{NT} = \begin{bmatrix} (P \otimes_s P^{-\top} S)Q_2 & (PX \otimes_s P^{-\top})(-\mathcal{A}_s^\top) \end{bmatrix}. \tag{5.44}$$

If P, $P^{-\top}$, $-\mathcal{A}_s^\top$, Q_2, PX and $P^{-\top}S$ are stored in QRAM. Then a

$$(\|M_{NT}\|_F, \mathcal{O}(\log n), \xi/(\|M_{NT}\|_F \kappa^2 \log^2 \tfrac{\kappa}{\xi}))$$

block-encoding of M_{NT} can be constructed in time $\widetilde{\mathcal{O}}_{n,\kappa,\frac{1}{\xi}}(1)$.

Proof. Carrying out the calculations shows that (5.44) corresponds to the Nesterov-Todd linear system. We construct the two following block-encodings:

$$M_1 = \begin{bmatrix} (P \otimes_s P^{-\top}S)Q_2 & 0 \end{bmatrix}$$
$$M_2 = \begin{bmatrix} 0 & (PX \otimes_s P^{-\top})(-\mathcal{A}_s^\top), \end{bmatrix}$$

using Propositions 5.18-5.19. We choose the precision of this step so that we obtain

$$(\|M_1\|_F, \mathcal{O}(\log n), \xi/(\|M_1\|_F \kappa^2 \log^2 \tfrac{\kappa}{\xi}))$$

and

$$(\|M_2\|_F, \mathcal{O}(\log n), \xi/(\|M_2\|_F \kappa^2 \log^2 \tfrac{\kappa}{\xi}))$$

block-encodings of M_1 and M_2, respectively, where κ refers to the condition number of (5.19), here and in the sequel. We add these two block-encodings together using Proposition 3.4, obtaining a

$$(\max\{\|M_1\|_F, \|M_2\|_F\}, \mathcal{O}(\log n), \xi/(\kappa^2 \log^2 \tfrac{\kappa}{\xi}))$$

block-encoding of (5.44), in time $\widetilde{\mathcal{O}}_{n,\kappa,\frac{1}{\xi}}(1)$. Since $\max\{\|M_1\|_F, \|M_2\|_F\} = \mathcal{O}(\|M_{NT}\|_F)$, we obtain the claimed result. $\qquad\square$

One should note that the construction of the Newton linear system provided in Proposition 5.20 is only one of the many possible ways to obtain a block-encoding of our Newton linear system while utilizing the Nesterov-Todd direction. Further, though the Nesterov-Todd direction is the search direction we study in this manuscript, note that our construction is written in general terms of P, thus this construction can be used for any suitable choice of P, such as the AHO ($P = I$) and HKM ($P = S^{1/2}$) directions. Note that by employing a factorization of the Newton linear system, we are able to theoretically improve the running time of implementing the block-encodings of the Newton linear system; which in turn allows us to improve the efficiency with which we solve the system. Next, we use this factorization to solve the Newton system.

Theorem 5.21. *Using the Nesterov-Todd direction $P = W^{-1/2}$, there is a quantum algorithm, which given $|\mathrm{svec}(R^c)\rangle$ and access to QRAM data structures encoding P, $P^{-\top}$, $-\mathcal{A}_s^\top$, Q_2, PX and $P^{-\top}S$, outputs a state ξ-close to $|\Delta z \circ \Delta y\rangle$ in time*

$$\widetilde{\mathcal{O}}_{n,\kappa,\frac{1}{\xi}}\left(\kappa \max\{\|M_1\|_F, \|M_2\|_F\}\right).$$

We can also output an estimate of $\|\Delta z \circ \Delta y\|$ with relative error δ by increasing the running time by a factor of $\frac{1}{\delta}$.

Proof. This is a direct consequence of Proposition 5.20 and Theorem 3.20. $\qquad\square$

5.6 Complexity

In this section, we formally analyze the worst case overall running times of the IF- and II-QIPMs. The per-iteration cost of either QIPM can be generally written as

$$T_{iter} = T_{NT} + T_{progress},$$

where T_{NT} is the quantum gate complexity of obtaining a classical solution to the Newton linear system (5.19), and $T_{progress}$ is the number of classical arithmetic operations required to prepare the next iteration. Thus, upon bounding T_{iter} for the IF-QIPM, one can obtain the worst-case overall complexity using the iteration bounds that were established in the previous section.

The following result establishes per-iteration cost of the IF-QIPM.

Proposition 5.22. *Each iteration k of the IF-QIPM in Algorithm 4 has a quantum gate complexity of*

$$\mathcal{O}\left(n^3\kappa^2 \cdot \operatorname{polylog}\left(n, \kappa, \frac{1}{\epsilon}\right)\right)$$

and requires $\mathcal{O}\left(n^4 \cdot \operatorname{polylog}\left(n, \kappa, \frac{1}{\epsilon}\right)\right)$ classical arithmetic operations.

Proof. In every iteration, we need to prepare and solve the NT Newton linear system (5.19), and obtain a classical estimate of the quantum state encoding its solution. In what follows, let ϱ be the maximum norm of a solution to the Newton linear system encountered during the algorithm.

Note that we use Algorithm 2 (Iterative refinement for the LSP) as a quantum subroutine that returns a classical description of the solution to the Newton system at each iteratre. Applying Theorem 4.5 with $n = d^2$ asserts that this step requires

$$\mathcal{O}\left(n^2\alpha\kappa\varrho \cdot \operatorname{polylog}\left(n, \kappa, \frac{1}{\epsilon}\right)\right)$$

accesses to the QRAM and $\mathcal{O}\left(n^4 \cdot \operatorname{polylog}\left(n, \kappa, \frac{1}{\epsilon}\right)\right)$ classical arithmetic operations. Here, α is the subnormalization factor used to block-encode the Nesterov-Todd coefficient matrix M_{NT} and ϱ is an upper bound on the norm of the solution to the Newton system. Hence, letting $\alpha = \|M_{NT}\|_F \leq n\|M_{NT}\|$ and noting the bound $\varrho = \mathcal{O}\left(\|M_{NT}^{-1}\|\right)$ provided in Proposition 5.15, thus, solving the Newton system at each iterate requires

$$\mathcal{O}\left(n^3\kappa^2 \cdot \operatorname{polylog}\left(n, \kappa, \frac{1}{\epsilon}\right)\right)$$

accesses to the QRAM and $\mathcal{O}\left(n^4 \cdot \text{polylog}\left(n, \kappa, \frac{1}{\epsilon}\right)\right)$ classical arithmetic operations.

To progress to the iteration, we must classically compute the primal and dual search directions ΔX and ΔS using our classical estimates of Δz and Δy using the nullspace representation provided in equations (5.18a)-(5.18b), and susbequently update the solution (X, y, S) to the SDO. We must also classically compute our scaling matrix P and its inverse. Computing ΔX and ΔS using (5.18a)-(5.18b) amounts to classically computing matrix-vector products in which the dimension is n^2, and hence requires $\mathcal{O}(n^4)$ arithmetic operations, while updating the solution is matrix addition involving $n \times n$ matrices, and can be accomplished using $\mathcal{O}(n^2)$ arithmetic operations. Finally, noting that all of the matrices involved in the computation of P have size $n \times n$, we compute P using $\mathcal{O}(n^\omega)$ arithmetic operations. Summarizing, we have

$$T^{IF}_{progress} = \mathcal{O}(n^4 + n^2 + n^\omega) = \mathcal{O}(n^4),$$

as computing ΔX and ΔS from our estimate of the solution to (5.19) is the dominant operation. Summing these costs, we arrive at the stated per-iteration cost of $T^{IF}_{iter} = T^{IF}_{NT} + T^{IF}_{progress}$, and the proof is complete. $\qquad\square$

We are now in a position to bound the overall running time of our IF-QIPM given in Algorithm 4, which is formalized in the following result.

Corollary 5.2. *A quantum implementation of Algorithm 4 with access to QRAM outputs an ϵ-optimal solution (X^*, y^*, S^*) to the primal-dual SDO pair (5.1)-(5.2) using at most*

$$\mathcal{O}\left(n^{3.5}\kappa^2 \cdot \text{polylog}\left(n, \kappa, \frac{1}{\epsilon}\right)\right)$$

QRAM accesses and $\mathcal{O}\left(n^{4.5} \cdot \text{polylog}\left(n, \kappa, \frac{1}{\epsilon}\right)\right)$ classical arithmetic operations.

Proof. By Corollary 5.1, the IF-QIPM requires $\mathcal{O}(\sqrt{n}\log(1/\epsilon))$ iterations to converge to an ϵ-optimal solution. Hence, applying Proposition 5.22, the algorithm converges to an ϵ-optimal solution after at most
$$\mathcal{O}\left(n^{4.5}\kappa^2 \cdot \text{polylog}\left(n, \kappa, \frac{1}{\epsilon}\right)\right)$$

QRAM accesses and $\mathcal{O}\left(n^{4.5} \cdot \text{polylog}\left(n, \kappa, \frac{1}{\epsilon}\right)\right)$ classical arithmetic operations. $\qquad\square$

Recall that LOPs are simply a special instance of SDO in which all of the matrices are diagonal. Hence, our algorithm can also be applied to these problems as well.

Corollary 5.3. *A quantum implementation of Algorithm 4 with access to QRAM outputs an ϵ-optimal solution $(X^*, y^*, S^*) = (\text{diag}(x^*), y^*, \text{diag}(s^*))$ to a Linear Optimization problem using at most*
$$\mathcal{O}\left(n^2\kappa^2 \cdot \text{polylog}\left(n, \kappa, \frac{1}{\epsilon}\right)\right)$$

accesses to the QRAM and $\mathcal{O}\left(n^{2.5} \cdot \text{polylog}\left(n, \kappa, \frac{1}{\epsilon}\right)\right)$ arithmetic operations.

Proof. The quantum gate complexity can be obtained upon repeating the analysis we used in the proof of Corollary 5.2, noting that for LO the dimension of the problem is $d = n$ instead of $d = n^2$. Then, recall that in the case of LO, all of the matrices are diagonal, and therefore no symmetrization is required. Hence, the only classical progression steps in this case are computing the primal and dual search directions from their nullspace representation, and subsequently updating the solution, and each of these steps require $\mathcal{O}(n^2)$ arithmetic operations. $\qquad\square$

Since our IF-QIPM utilizes a hybrid quantum-classical scheme, one can trivially obtain a fully classical IF-IPM by replacing the quantum subroutine for solving the Newton linear system with an inexact classical linear systems algorithm. Note that the system we obtain in (5.19) using the nullspace representation is not guaranteed to be positive semidefinite, so we cannot directly apply the Conjugate Gradient Method (CG) to solve it. For $M \in \mathbb{R}^{d \times d}$ and $v \in \mathbb{R}^d$ with $M \notin \mathcal{S}^n_+$, one could first symmetrize the system $Mu = v$ as

$$M^2 u = Mv,$$

and subsequently apply CG, in which case the cost of this step would be

$$\mathcal{O}\left(d^\omega + d^2 \sqrt{\kappa_M^2}\,\log(1/\xi)\right) = \mathcal{O}\left(d^\omega + d^2 \kappa_M \log(1/\xi)\right),$$

because we need to carry out matrix-matrix multiplication, and the condition number of the resulting coefficient matrix is κ_M^2. Alternatively, we could employ one of the generalizations of CG such as the Generalized Minimum Residual Method (GMRES), however, not only is the coefficient matrix we employ indefinite, but it is nonsymmetric and dense.[2]: applying GMRES to solve a dense, nonsymmetric and indefinite system of dimension d to precision ξ leads to a complexity of $\mathcal{O}\left(d^3 \log(1/\xi)\right)$. In our case $d = n^2$, so classically performing this step will require time $\mathcal{O}(n^{2\omega} + n^4 \kappa \log(1/\xi))$ if we symmetrize and use CG, or time $\mathcal{O}\left(n^6 \log(1/\xi)\right)$ using GMRES.

This outcome is undesirable because it would lead to a slower algorithm than what already exists in the classical literature. Alternatively, one can use a polynomial approximation q, of $1/u$ on the interval $[1/\kappa_M^2, 1]$. The solution would then be $q(M^2)Mv$, which is approximately $M^{-2}Mv = M^{-1}v$. Such a polynomial q with degree roughly $\kappa_M \log(1/\xi)$ exists (see, e.g., [Saa03, Section 6.11]), and the approximate solution would be $q(M^2)Mv$, which is close to $M^{-1}v$ whenever the linear system $Mu = v$ has a solution. Crucially, $q(M^2)Mv$ can be obtained using $2 \cdot \deg(q) + 1$ matrix-vector products, and does not require computing (or even writing down) the matrix M^2, thus avoiding the $n^{2\omega}$ present term in the complexity associated with symmetrization and subsequently using CG. Rather, in our case the complexity becomes

$$\mathcal{O}(d^2 \sqrt{\kappa_M}\,\log(1/\xi)) = \mathcal{O}\left(\left(n^2\right)^2 \sqrt{\kappa^2}\,\log(1/\xi)\right) = \mathcal{O}\left(n^4 \kappa \log(1/\xi)\right).$$

For an explicit discussion of this method, we refer the reader to Section 16.5 of the monograph

---

[2] We do not review CG or GMRES in detail, and instead refer the reader to Chapter 6 of [Saa03].

[Vis13].

This argument is the basis for the next result, which bounds the running time of the classical IF-IPM for SDO.

Corollary 5.4. *A classical implementation of Algorithm 4 obtains an ϵ-optimal solution (X^*, y^*, S^*) to the primal-dual SDO pair (5.1)-(5.2) with overall complexity*

$$\mathcal{O}\left(n^{4.5}\kappa \cdot \mathrm{polylog}\left(\kappa, \frac{1}{\epsilon}\right)\right).$$

Proof. Simply note that applying [Vis13, Theorem 16.6] with $d = n^2$ one has

$$T_{NT}^{IF} = \mathcal{O}\left(n^4\kappa \cdot \mathrm{polylog}\left(\kappa, \frac{1}{\epsilon}\right)\right),$$

and we still have $T_{progress}^{IF} = \mathcal{O}(n^4)$.

Just as in the case of the IF-QIPM, the linear system solver we use for our IF-IPM takes the condition number of the Newton linear system as input. Using the verification scheme described in the proof of Theorem 4.5 at each iterate implies that we might need to repeat these steps $\log(\kappa)$ times to ensure the solution we obtain is accepted. The stated result then follows upon accounting for the iteration bound provided by Corollary 5.1. $\square$

Corollary 5.5. *A classical implementation of Algorithm 4 obtains an ϵ-optimal solution*

$$(X^*, y^*, S^*) = (\mathrm{diag}(x^*), y^*, \mathrm{diag}(s^*))$$

to a Linear Optimization problem in time

$$\mathcal{O}\left(n^{2.5}\kappa \cdot \mathrm{polylog}\left(\kappa, \frac{1}{\epsilon}\right)\right).$$

5.6.1 Comparison with existing algorithms for SDO

Before proceeding further, we remind the reader of a number of key differences between (Q)IPMs and (Q)MWU algorithms. The output of (Q)IPMs is a classical primal-dual pair (X, y, S) such that

$$\mathrm{tr}\,(A_i X) = b_i \quad \forall i \in [m], \quad X \succ 0$$

$$\sum_{i \in [m]} y_i A_i + S = C \quad S \succeq 0$$

with

$$\frac{\mathrm{tr}\,(XS)}{n} \leq \epsilon.$$

Conversely, primal-dual (Q)MWUs output a dual solution $y \in \mathbb{R}^m$ such that setting

$$X = R \cdot \frac{\exp\left(-\sum_{i=1}^m y_i A_i\right)}{\operatorname{tr}\left(\exp\left(-\sum_{i=1}^m y_i A_i\right)\right)}$$

the primal-dual pair (X, y) satisfies

$$\left|\operatorname{tr}(A_i X) - b_i\right| \leq \epsilon_{\text{abs}} \quad \forall i \in [m], \quad X \succ 0$$
$$S = C - \sum_{i \in [m]} y_i A_i \succeq -\epsilon_{\text{abs}} I$$

and the objective value attained by this solution is $\text{OPT} \in [\varsigma - \mathcal{O}(\epsilon_{\text{abs}}), \varsigma + \mathcal{O}(\epsilon_{\text{abs}})]$, where ς is a bound on the optimal objective value determined using binary search. Observe that this is another distinction between (Q)IPMs and (Q)MWUs; the output of our QIPMs always satisfies primal and dual feasibility exactly, whereas the solution obtained by (Q)MWU methods satisfies primal and dual feasibility approximately.

The QMMWUs (specifically, all papers by Brandao et al. [BKL+19, BKF22, BS17], as well as the papers by van Apeldoorn et al. [vAG19, vAGGdW20] also assume that the SDO problem data is normalized:

$$\|A_1\|, \ldots, \|A_m\|, \|C\| \leq 1.$$

Our QIPMs do not require this assumption. For the QIPMs presented in this work, only the Newton linear system must be normalized in each iteration (so we can apply a QLSA to solve it), and this normalization is accounted for in our accuracy requirements and the subsequent running time analysis. (For classical MMWUs, the choice of the normalization can vary depending on the papers; the analysis of the classical algorithm in [vAGGdW20] assumes the above normalization.)

Assuming the data is normalized with respect to the operator norm means that for more general problems (i.e., without normalized data) the error scales as $\alpha \epsilon_{\text{abs}}$, where

$$\alpha = \max\left\{\|A_1\|, \ldots, \|A_m\|, \|C\|\right\}.$$

Thus, for MMWUs one would lose a factor α in the precision. Note that being able to compute α to normalize the data as $\alpha^{-1} A_1, \ldots, \alpha^{-1} A_m, \alpha^{-1} C$ could require classical access to these matrices, and an additional $\mathcal{O}(mns)$ cost in the running time to compute α.

Table 5.2 presents the running time of our algorithms alongside well-known algorithms for solving SDOPs. For ease of comparison, we assume that $m = \mathcal{O}(n^2)$ and that the problem is fully dense (i.e., $s = n$). It can be readily seen that the II-QIPM is worse in every parameter when compared to the other algorithms. Conversely, the IF-QIPM obtains a speedup with respect to the dimension n over each of the listed classical solution methodologies. Yet, the IF-QIPM exhibits a linear dependence on the condition number bound κ_T, so it is inconclusive whether there is an overall speedup. worse running time overall.

References	Method	Runtime
[Mon98, NT97, NT98]	IPM	$\mathcal{O}_{n,\frac{1}{\epsilon}}\left(n^{6.5}\right)$
[JKL$^+$20]	IPM	$\widetilde{\mathcal{O}}_{\frac{1}{\epsilon}}\left(n^{5.246}\right)$
[JLSW20, LSW15]	CPM	$\widetilde{\mathcal{O}}_{n,R,\frac{1}{\epsilon}}\left(n^{6}\right)$
[vAG19]	QMWU	$\widetilde{\mathcal{O}}_{n,\frac{1}{\epsilon_{\mathrm{abs}}}}\left(\left(n^2+n^{1.5}\frac{Rr}{\epsilon_{\mathrm{abs}}}\right)\left(\frac{Rr}{\epsilon_{\mathrm{abs}}}\right)^4\right)$
This work (Quantum)	IF-QIPM	$\widetilde{\mathcal{O}}_{n,\kappa,\frac{1}{\epsilon}}\left(\sqrt{n}\left(n^3\kappa^2+n^4\right)\right)$
This work (Classical)	IF-IPM	$\widetilde{\mathcal{O}}_{\kappa,\frac{1}{\epsilon}}\left(n^{4.5}\kappa\right)$

Table 5.1: Total running times for classical and quantum algorithms to solve (5.1)-(5.2) with $m = \mathcal{O}(n^2)$.

When compared to the QMWU algorithm, it can be observed that the QIPM achieves a favorable dependence on the error. Although the QMWU is faster with respect to dimension, the QMWU has polynomial dependence on the trace bound R, and for certain applications this may negate any speedup, such as the SDO approximation of Quadratic Unconstrained Binary Optimization (QUBO) problems, i.e., problems of the form

$$\max_{x\in\{-1,1\}^n} x^\top Cx. \tag{5.45}$$

QUBOs play an important role in quantum information sciences, and the resulting SDO approximation is given by

$$\max_{X\in\mathcal{S}^n}\left\{\mathrm{tr}(CX) : \mathrm{diag}(X) = e, X \succeq 0\right\}, \tag{5.46}$$

where e is the all ones vector of dimension n. Hence, any optimal solution of (5.48) satisfies $\mathrm{tr}(X) = n$, suggesting that for these instances the worst case running time of the QMWU is

$$\widetilde{\mathcal{O}}_{n,\frac{1}{\epsilon_{\mathrm{abs}}}}\left(n^{6.5}\left(\frac{r}{\epsilon_{\mathrm{abs}}}\right)^5\right),$$

as $R = n$ and $m = n$ for (5.48). Thus, the IF-QIPM could outperform the QMWU for this class of SDOPs, provided that $\kappa = \mathcal{O}\left(nr^{2.5}\left(\frac{1}{\epsilon}\right)^2\right)$.

For SDOPs with diagonal constraints such as (5.48), the Arora-Kale method can be specialized (see, e.g., [LP20]) to obtain an $\tilde{\mathcal{O}}(ns\epsilon^{-3.5})$ algorithm. Note that to achieve this running time, the algorithm avoids explicitly computing or reporting the primal solution X, and instead makes use of techniques to estimate its diagonal entries and trace inner products of the form $\mathrm{tr}(AX)$. As an alternative, these methods report a "gradient" $G \in \mathcal{S}^n$ such that $X = W\exp(G)W$ for a diagonal matrix W. While the algorithm of Lee and Padmanabhan [LP20] offers speedups with respect to the dimension, its running time is exponentially slower than classical IPMs with respect to the inverse precision.

To obtain an additive ϵ-convergence using the algorithm found in [LP20], the error parameter would have to be set to

$$\tilde{\epsilon} = \frac{\epsilon}{\|C\|_{\ell_1}}.$$

When C is dense and $\|C\|_{\ell_1} = \mathcal{O}(n)$, the running time of their algorithm therefore scales as

$$\mathcal{O}(n^2(\tilde{\epsilon})^{-3.5}) = \mathcal{O}\left(n^2\|C\|_{\ell_1}^{3.5}\epsilon^{-3.5}\right).$$

For our QIPM to reach duality gap ϵ would require complexity

$$\tilde{\mathcal{O}}_{n,\kappa,\frac{1}{\epsilon}}\left(\sqrt{n}\left(n^3\kappa^2 + n^4\right)\right).$$

However, one could solve the QUBO problem to ϵ-optimality using the IPM from Jiang et al. [JKL$^+$20] in time $\mathcal{O}(n^{\omega+0.5}\log(n/\epsilon))$, or even our classical IF-IPM with overall running time $\tilde{\mathcal{O}}_{\kappa,\frac{n}{\epsilon}}\left(n^{4.5}\kappa\right)$.

5.7 Iterative Refinement for SDO using QIPMs

In the classical IPM literature for solving LO problems, several studies address the numerical analysis of reaching an exact solution, such as the Iterative Refinement scheme by [MAF$^+$22] and the Rational Reconstruction method by [GS20]. In this section we show how to use Iterative Refinement as a general procedure to improve the precision of solution for quantum Interior Point Methods for Semidefinite Optimization.

The basic idea is to again solve a sequence of related problems, only this time the solutions improve the accuracy to which we solve an SDO problem rather than a linear system of equations. Suppose we have access to an oracle O_{FO} which that returns feasible solutions to a given SDO problem, such as the IF-QIPM provided in Algorithm 4. Using a feasible oracle, we can define an Iterative Refinement scheme for SDO with QIPMs as outlined in Algorithm 5.

The algorithm takes as input the input data for the primal dual pair (5.1)-(5.2), and two error tolerances; ϵ, the *fixed* precision to which each oracle call is made, and $\tilde{\epsilon}$, the desired normalized duality gap of the final solution. In each iteration we query the IF-QIPM with the SDO problem data, which then reports the refining solution $(\bar{X}, \bar{y})$. From here the solution to the SDO problem is updated, as well as the refining problem data. We then compute the normalized duality gap associated with the current solution $\bar{\mu}$. If $\bar{\mu}$ has been reduced to or below $\tilde{\epsilon}$, the algorithm terminates and reports (X, y) as a $\tilde{\epsilon}$-optimal solution to (5.1)-(5.2). Otherwise, we proceed to the next iteration.

The next result explicitly defines the primal and dual refining problems.

Definition 5.1. *Consider the primal-dual SDO pair (5.1)-(5.2) and assume that the IPC is satisfied such that Strong Duality holds. Let (X, y, S) be the current solution to (5.1)-(5.2). For any $\overline{X} \in \mathcal{S}^n$ and $\bar{y} \in \mathbb{R}^m$ and scaling factor, $\eta \geq 1$ consider the primal refining problem is defined as*

$$\min\left\{\operatorname{tr}\left(\eta\overline{C}\overline{X}\right) : \operatorname{tr}(A_i\overline{X}) = \eta\bar{b}_i \text{ for } i \in [m] \text{ and } \overline{X} \succeq -\eta X\right\}, \tag{$\bar{P}$}$$

Algorithm 5 Iterative Refinement for SDO using QIPMs

Input: Problem data $A_1, \dots, A_m, C \in \mathcal{S}^n$, $b \in \mathbb{R}^m$, error tolerances $0 < \tilde{\epsilon} \ll \epsilon < 1$. Strictly feasible point for (5.1)-(5.2) $(X^{(0)}, y^{(0)}, S^{(0)})$

Output: A $\tilde{\epsilon}$-optimal primal-dual solution (X, y, S) to the SDO problem $(A_1, \dots, A_m, b, C)$

Initialize: $\eta^{(1)} \leftarrow 1$, $\bar{\mu}^{(0)} \leftarrow 1$, $k \leftarrow 1$, $\bar{b} \leftarrow b$, $\bar{C} \leftarrow C$

$(X^{(1)}, y^{(1)}) \leftarrow$ **solve** (5.1)-(5.2) to precision ϵ using Algorithm 4

Compute residuals

$$\bar{b}_i \leftarrow b_i - \text{tr}(A_i X^{(1)}) \text{ for } i \in [m], \quad \bar{C} \leftarrow C - \sum_{i \in [m]} A_i^\top y^{(1)}$$

Compute normalized duality gap at updated solution

$$\bar{\mu}^{(1)} \leftarrow \frac{\text{tr}\left[X^{(1)}\left(C - \sum_{i \in [m]} A_i^\top y^{(1)}\right)\right]}{n}$$

Update scaling factor

$$\eta^{(2)} \leftarrow \frac{1}{\bar{\mu}^{(1)}}$$

$k \leftarrow k + 1$

while $\bar{\mu} > \tilde{\epsilon}$ **do**

    1. $(\overline{X}, \bar{y}) \leftarrow$ **solve** $(\bar{P})$-$(\bar{D})$ with input data $(A_1, \dots, A_m, \eta^{(k)}\bar{b}, \eta^{(k)}\bar{C})$ to precision ϵ using Algorithm 4

    2. Update solution

$$X^{(k+1)} \leftarrow X^{(k)} + \frac{1}{\eta^{(k)}}\overline{X}, \quad y^{(k+1)} \leftarrow y^{(k)} + \frac{1}{\eta^{(k)}}\bar{y}$$

    3. Update refining problem data

$$\bar{b}_i^{(k+1)} \leftarrow b_i - \text{tr}(A_i X^{(k+1)}) \text{ for } i \in [m], \quad \bar{C}^{(k+1)} \leftarrow C - \sum_{i=1}^m y_i^{(k+1)} A_i$$

    4. Compute residual $\bar{\mu}^{(k+1)} = \frac{\text{tr}\left(X^{(k+1)}\bar{C}\right)}{n}$

    5. Update scaling factor $\eta^{(k+1)} = \frac{1}{\bar{\mu}^{(k+1)}}$

    6. $k \leftarrow k + 1$

end

---

and its dual problem

$$\max \left\{ \eta \bar{b}^T \bar{y} - \eta \operatorname{tr} \left(X \left[\overline{C} - \sum_{i=1}^{m} \bar{y}_i A_i \right] \right) : \overline{C} - \sum_{i=1}^{m} \bar{y}_i A_i \succeq 0 \right\}, \tag{$\bar{D}$}$$

where $\overline{C} = C - \sum_{i=1}^{m} y_i A_i$ and $\bar{b}_i = b_i - \operatorname{tr}(A_i X)$.

The following result establishes that the sequence of iterates generated by Algorithm 4 are increasingly accurate solutions to the primal-dual pair (5.1)-(5.2).

Theorem 5.23. *Let $\left(X^{(k)}, y^{(k)} \right)$ be the current solution, and $\left(\overline{X}^{(k)}, \bar{y}^{(k)} \right)$ be an ϵ-optimal solution to the refining problem at iteration $k \geq 0$ of Algorithm 5. Then, $\left(X^{(k+1)}, y^{(k+1)} \right)$ is an ϵ^{k+1}-optimal solution to (5.1)-(5.2).*

Proof. Recall, by definition

$$\eta^{(k)} \bar{b}_i^{(k)} = \eta^{(k)} \left[b_i - \operatorname{tr} \left(A_i X^{(k)} \right) \right]$$

$$\eta^{(k)} \overline{C}^{(k)} = \eta^{(k)} \left(C - \sum_{i=1}^{m} y_i^{(k)} A_i \right).$$

Any solution $\left(\overline{X}^{(k)}, \bar{y}^{(k)} \right)$ to $(\bar{P})$-$(\bar{D})$ satisfies

$$\operatorname{tr} \left(A_i \overline{X}^{(k)} \right) = \eta^{(k)} \bar{b}_i, \quad X^{(k)} + \frac{1}{\eta^{(k)}} \overline{X}^{(k)} \succeq 0,$$

$$\eta^{(k)} \overline{C}^{(k)} - \sum_{i=1}^{m} \bar{y}_i^{(k)} A_i \succeq 0,$$

$$\frac{\operatorname{tr} \left[\left(\eta^{(k)} X^{(k)} + \overline{X}^{(k)} \right) \left(\eta^{(k)} \bar{C}^{(k)} - \sum_{i=0}^{m} \bar{y}_i^{(k)} A_i \right) \right]}{n} \leq \epsilon,$$

and hence $X^{(k+1)} = X^{(k)} + \frac{1}{\eta^{(k)}} \overline{X}^{(k)} \succeq 0$.

Next, note that for any $k \geq 0$, the updated solution will satisfy primal and dual feasibility. For all $i \in [m]$, we have

$$\operatorname{tr} \left(A_i X^{(k+1)} \right) = \operatorname{tr} \left(A_i \left(X^{(k)} + \frac{1}{\eta^{(k)}} \overline{X}^{(k)} \right) \right) = \operatorname{tr} \left(A_i X^{(k)} \right) + \frac{1}{\eta^{(k)}} \eta^{(k)} \bar{b}_i$$

$$= \operatorname{tr} \left(A_i X^{(k)} \right) + \left[b_i - \operatorname{tr} \left(A_i X^{(k)} \right) \right] = b_i.$$

Similarly,

$$C - \sum_{i=1}^{m} A_i^\top y_i^{(k+1)} = C - \sum_{i=1}^{m} A_i^\top \left(y_i^{(k)} + \frac{1}{\eta^{(k)}} \bar{y}_i^{(k)} \right)$$

$$= \frac{1}{\eta^{(k)}} \left(\eta^{(k)} \left(C - \sum_{i=1}^{m} A_i^\top y_i^{(k)} \right) - \sum_{i=1}^{m} A_i^\top \bar{y}^{(k)} \right)$$

$$= \frac{1}{\eta^{(k)}} \left(\eta^{(k)} \overline{C}^{(k)} - \sum_{i=1}^{m} A_i^\top \bar{y}_i^{(k)} \right) \succeq 0.$$

Moreover, the normalized duality gap is given by

$$\frac{\operatorname{tr}\left[X^{(k+1)} \left(C - \sum_{i=1}^{m} y_i^{(k+1)} A_i \right) \right]}{n} = \frac{\operatorname{tr}\left[\left(X^{(k)} + \frac{1}{\eta^{(k)}} \overline{X}^{(k)} \right) \left(C - \sum_{i=1}^{m} \left(y_i^{(k)} + \frac{1}{\eta^{(k)}} \bar{y}^{(k)} \right) A_i \right) \right]}{n}$$

$$= \frac{\operatorname{tr}\left[\left(\frac{1}{\eta^{(k)}} \left(\eta^{(k)} X^{(k)} + \overline{X}^{(k)} \right) \right) \left(\frac{1}{\eta^{(k)}} \left(\eta^{(k)} \overline{C}^{(k)} - \sum_{i=1}^{m} A_i^\top \bar{y}_i^{(k)} \right) \right) \right]}{n}$$

$$= \frac{1}{(\eta^{(k)})^2} \frac{\operatorname{tr}\left[\left(\eta^{(k)} X^{(k)} + \overline{X}^{(k)} \right) \left(\eta^{(k)} \bar{C}^{(k)} - \sum_{i=1}^{m} \bar{y}_i^{(k)} A_i \right) \right]}{n}$$

$$\leq \frac{\epsilon}{\eta^{(k)}}.$$

Therefore, our claim holds as long as $\eta^{(k)} \geq \frac{1}{\epsilon^k}$, which we will prove using a standard induction argument.

Observe that $\eta^{(k)} \geq \frac{1}{\epsilon^k}$ trivially holds for $k = 0$ by initializing $\eta^{(0)} = 1 = \frac{1}{\epsilon^0}$. Applying the induction hypothesis, we now assume that $\eta^{(\ell)} \geq \frac{1}{\epsilon^\ell}$ is true for all $\ell = 1, \dots, k$. Then, noting that the normalized duality gap of the solution at iteration $k + 1$ is $\frac{\epsilon}{\eta^{(k)}}$, it follows that

$$\eta^{(k+1)} = \frac{1}{\frac{\operatorname{tr}\left[X^{(k+1)} \left(C - \sum_{i=1}^{m} y_i^{(k+1)} A_i \right) \right]}{n}} \geq \frac{\eta^{(k)}}{\epsilon} \geq \frac{\frac{1}{\epsilon^k}}{\epsilon} = \frac{1}{\epsilon^{k+1}}.$$

Therefore, $\eta^{(k)} \geq \frac{1}{\epsilon^k}$ holds for all $k \geq 0$, and the normalized duality gap at iteration $k + 1$ of Algorithm 5 is at most

$$\frac{\epsilon}{\eta^{(k)}} \leq \frac{\epsilon}{\frac{1}{\epsilon^k}} = \epsilon^{k+1},$$

which completes the proof. $\square$

Corollary 5.6. *Algorithm 5 obtains a $\tilde{\epsilon}$-precise solution to the primal-dual SDO pair (5.1)-(5.2) in at most*

$$\mathcal{O}\left(\log\left(\frac{1}{\tilde{\epsilon}} \right) \right)$$

iterations.

Proof. The result is an immediate consequence of Theorem 5.23. $\square$

5.7.1 Choosing a starting point for the refining iterates

Now in any feasible IPM scheme it is necessary to assume the existence of a strictly feasible point $\left(X^{(0)}, y^{(0)}, S^{(0)}\right)$ such that the IPC is satisfied. As we have stressed, this can be made without loss of generality because we can always use the Self-Dual Embedding model, for which there exists a trivial interior feasible point. However, for both theory and practice, it is worthwhile to show how one can easily obtain an initial interior point for each refining problem.

The following result generalizes [MFWT21, Theorem 6.4] from LO to SDO.

Theorem 5.24 (Adapted from Theorem 6.4 in [MFWT21]). *Let* $(X^{(0)}, y^{(0)}, S^{(0)}) \in \mathcal{S}^n_{++} \times \mathbb{R}^m \times \mathcal{S}^n_{++}$ *be an interior feasible solution to the original primal-dual SDO pair (5.1)-(5.2), and let* $\mu^{(0)}$ *be the noramlized duality gap at this solution. Let* $\left(X^{(k)}, y^{(k)}, S^{(k)}\right) \in \mathcal{S}^n_+ \times \mathbb{R}^m \times \mathcal{S}^n_+$ *and* $\eta^{(k)}$ *be the solution and scaling factor at iterate* k *of Algorithm ??, respectively. Then, the point*

$$\left(\eta^{(k)} \left(X - X^{(k)}\right), \eta^{(k)} \left(y - y^{(k)}\right), \eta^{(k)} S^{(0)}\right) \in \mathcal{S}^n \times \mathbb{R}^m \times \mathcal{S}^n_{++}$$

is a strictly feasible solution to the refining problem at iteration $k+1$. *Moreover, the normalized duality gap at this starting point is* $\left(\eta^{(k)}\right)^2 \mu^{(0)}$.

Proof. First, note that for all $i \in [m]$, we have

$$\operatorname{tr}\left(A_i \left[\eta^{(k)}\left(X^{(0)} - X^{(k)}\right)\right]\right) = \eta^{(k)}\left(\operatorname{tr}\left(A_i X^{(0)}\right) - \operatorname{tr}\left(A_i X^{(k)}\right)\right) = \eta^{(k)}(b_i - b_i) = 0.$$

Similarly,

$$\sum_{i \in [m]} A_i^\top \left[\eta^{(k)}\left(y - y^{(k)}\right)\right] + \eta^{(k)} S^{(0)} = \eta^{(k)} \left(\sum_{i \in [m]} A_i^\top y - \sum_{i \in [m]} A_i^\top y^{(k)} + S^{(0)}\right)$$

$$= \eta^{(k)} \left(C - \sum_{i \in [m]} A_i^\top y^{(k)}\right)$$

$$= \eta^{(k)} \overline{C}^{(k)}.$$

Now, for any strictly feasible solution $\bar{X}$ to the refining problem, we must have

$$X^{(k)} + \frac{1}{\eta^{(k)}} \bar{X}^{(k)} \succ 0,$$

which can be equivalently expressed as

$$X^{(k)} + \frac{1}{\eta^{(k)}} \bar{X} \succ 0 \iff \bar{X} \succ -\eta^{(k)} X^{(k)}.$$

Noting that $X^{(0)} \succ 0$ by definition, the point $\eta^{(k)}\left(X - X^{(k)}\right)$ is clearly strictly feasible, as

$$\eta^{(k)}\left(X^{(0)} - X^{(k)}\right) = \eta^{(k)} X^{(0)} - \eta^{(k)} X^{(k)} \succ -\eta^{(k)} X^{(k)}.$$

We must also have $\eta^{(k)} S^{(0)} \succ 0$ because $S^{(0)} \succ 0$ and $\eta^{(k)}$ is strictly positive for all k.

Therefore,

$$\left(\eta^{(k)} \left(X^{(0)} - X^{(k)} \right), \eta^{(k)} \left(y - y^{(k)} \right), \eta^{(k)} S^{(0)} \right) \in \mathcal{S}^n \times \mathbb{R}^m \times \mathcal{S}^n_{++}$$

is a strictly feasible solution to the refining problem at iteration k, and is a valid initial point for the QIPM. This completes the proof up to trivial computation. $\qquad\square$

5.7.2 The relationship between the condition number and optimality in IPMs

To motivate the idea at hand, let us consider the case of linear optimization. We reiterate here that this is illustrates precisely why dependence on the condition number is problematic for QIPMs; we seek to follow the central path as $\mu \to 0$, which of course implies κ will almost always grow to infinity as we reach optimality, regardless of the problem class (linear, second-order conic, semidefinite), or whether the problem data $\mathcal{A}$ itself is well-conditioned.

To bound the condition number of our coefficient matrix at the k-th iterate

$$M^{(k)} = \begin{bmatrix} -\mathcal{F}_s \mathcal{A}_s^\top & \mathcal{E}_s Q_2 \end{bmatrix},$$

the authors in [MFWT21] compute the condition number of $(M^{(k)})^\top M^{(k)}$ which can be written as follows

$$(M^{(k)})^\top M^{(k)} = \begin{bmatrix} \mathcal{A}_s & 0 \\ 0 & Q_2^\top \end{bmatrix} \begin{bmatrix} \mathcal{F}_s^2 & -\mathcal{F}_s^\top \mathcal{E}_s \\ -\mathcal{E}_s^\top \mathcal{F}_s & \mathcal{E}_s^2 \end{bmatrix} \begin{bmatrix} \mathcal{A}^\top & 0 \\ 0 & Q_2 \end{bmatrix}.$$

We then have the following result from [MFWT21].

Lemma 5.25 ([MFWT21]). *For any full row-rank matrix $T \in \mathbb{R}^{m \times n}$, and any symmetric positive definite matrix $D \in \mathbb{R}^{n \times n}$, we have*

$$\kappa(TDT^\top) \leq \kappa(D)\kappa(TT^\top).$$

Let

$$D^{(k)} = \begin{bmatrix} \mathcal{F}_s^2 & -\mathcal{F}_s^\top \mathcal{E}_s \\ -\mathcal{E}_s^\top \mathcal{F}_s & \mathcal{E}_s^2 \end{bmatrix}, T = \begin{bmatrix} \mathcal{A}^\top & 0 \\ 0 & Q_2 \end{bmatrix}.$$

Then, the following result from [MFWT21] bounds the factor that involves the symmetrization terms.

Lemma 5.26 ([MFWT21]). *Suppose $(X, y, S) \in \mathcal{P}^0 \times \mathcal{D}^0$. Then, the condition number of $D^{(k)}$ is $\mathcal{O}(\frac{1}{\sqrt{\mu^{(k)}}})$.*

This leads to the main result, which provides a bound for the condition number of the OSS representation of the Newton linear system.

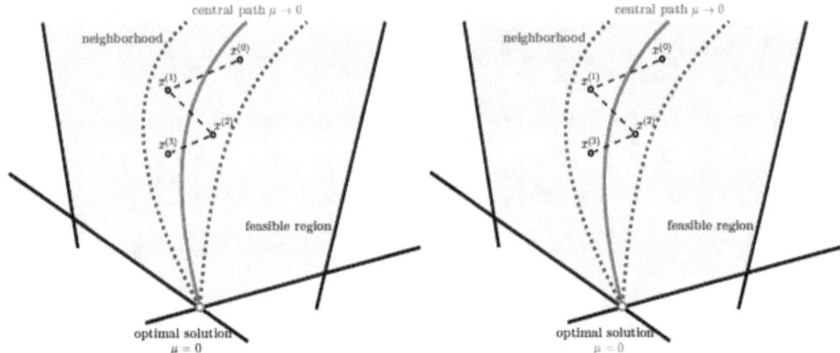

Figure 5.1: Conceptualization of an interior point method with "outer" iterative refinement

Theorem 5.27. *Suppose $(X, y, S) \in \mathcal{P}^0 \times \mathcal{D}^0$. Then, the condition number of (5.19) is bounded by $\mathcal{O}(\frac{\kappa_T}{\mu^{(k)}})$, where κ_T is the condition number of the matrix T.*

Hence, if we are using our QIPM to solve the SDO problem at hand to ϵ-optimality, $\mu > \epsilon$ holds for every iteration of the QIPM, and we have

$$\mathcal{O}\left(\kappa_T \frac{1}{\sqrt{\mu}}\right) = \mathcal{O}\left(\kappa_T \frac{1}{\sqrt{\epsilon}}\right).$$

In particular, in every oracle call made during Algorithm 5, we solve an SDO problem using fixed precision. Accordingly, the condition number of the Newton system coefficient matrix is bounded in each iterate as follows

$$\kappa = \mathcal{O}(\kappa_T).$$

Noting these facts, we are in a position to state the main result of this chapter.

Theorem 5.28. *Algorithm 5 obtains a $\tilde{\epsilon}$-precise solution (X^*, y^*, S^*) to the primal-dual SDO pair (5.1)-(5.2) in time*

$$\mathcal{O}\left(\sqrt{n}\left(n^2 \kappa_T^2 + n^4\right) \cdot \mathrm{polylog}\left(n, \kappa_U, \frac{1}{\tilde{\epsilon}}\right)\right).$$

Proof. The result follows upon combining Corollaries 5.2 and 5.6 with Theorem 5.27. $\square$

5.7.3 Comparison with existing algorithms for SDO

Before proceeding further, we remind the reader of a number of key differences between (Q)IPMs and (Q)MWU algorithms. The output of (Q)IPMs is a classical primal-dual pair (X, y, S) such that

$$\mathrm{tr}\,(A_i X) = b_i \quad \forall i \in [m], \quad X \succ 0$$

$$\sum_{i \in [m]} y_i A_i + S = C \quad S \succeq 0$$

with

$$\frac{\operatorname{tr}(Xs)}{n} \leq \epsilon.$$

Conversely, primal-dual (Q)MWUs output a dual solution $y \in \mathbb{R}^m$ such that setting

$$X = R \cdot \frac{\exp\left(-\sum_{i=1}^m y_i A_i\right)}{\operatorname{tr}\left(\exp\left(-\sum_{i=1}^m y_i A_i\right)\right)}$$

the primal-dual pair (X, y) satisfies

$$\left|\operatorname{tr}(A_i X) - b_i\right| \leq \epsilon_{\text{abs}} \quad \forall i \in [m], \quad X \succ 0$$

$$S = C - \sum_{i \in [m]} y_i A_i \succeq -\epsilon_{\text{abs}} I$$

and the objective value attained by this solution is OPT $\in [\varsigma - \mathcal{O}(\epsilon_{\text{abs}}), \varsigma + \mathcal{O}(\epsilon_{\text{abs}})]$, where ς is a bound on the optimal objective value determined using binary search. Observe that this is another distinction between (Q)IPMs and (Q)MWUs; the output of our QIPMs always satisfies primal and dual feasibility exactly, whereas the solution obtained by (Q)MWU methods satisfies primal and dual feasibility approximately.

The QMMWUs (specifically, all papers by Brandao et al. [BKL$^+$19, BKF22, BS17], as well as the papers by van Apeldoorn et al. [vAG19, vAGGdW20] also assume that the SDO problem data is normalized:

$$\|A_1\|, \ldots, \|A_m\|, \|C\| \leq 1.$$

Our QIPMs do not require this assumption. For the QIPMs presented in this work, only the Newton linear system must be normalized in each iteration (so we can apply a QLSA to solve it), and this normalization is accounted for in our accuracy requirements and the subsequent running time analysis. (For classical MMWUs, the choice of the normalization can vary depending on the papers; the analysis of the classical algorithm in [vAGGdW20] assumes the above normalization.)

Assuming the data is normalized with respect to the operator norm means that for more general problems (i.e., without normalized data) the error scales as $\alpha\epsilon_{\text{abs}}$, where

$$\alpha = \max\left\{\|A_1\|, \ldots, \|A_m\|, \|C\|\right\}.$$

Thus, for MMWUs one would lose a factor α in the precision. Note that being able to compute α to normalize the data as $\alpha^{-1}A_1, \ldots, \alpha^{-1}A_m, \alpha^{-1}C$ could require classical access to these matrices, and an additional $\mathcal{O}(mns)$ cost in the running time to compute α.

Table 5.2 presents the running time of our algorithms alongside well-known algorithms for solving SDOPs. For ease of comparison, we assume that $m = \mathcal{O}(n^2)$ and that the problem is fully dense (i.e., $s = n$). It can be readily seen that the II-QIPM is worse in every parameter when compared to the other algorithms. Conversely, the IF-QIPM obtains a speedup with respect to the dimension n over each of the listed classical solution methodologies. Yet, the IF-QIPM exhibits a linear dependence on the condition number bound κ_T, so it is inconclusive whether there is an overall speedup. worse running time overall.

References	Method	Runtime
[Mon98, NT97, NT98]	IPM	$\widetilde{\mathcal{O}}_{n,\frac{1}{\epsilon}}\left(n^{6.5}\right)$
[JKL$^+$20]	IPM	$\widetilde{\mathcal{O}}_{\frac{1}{\epsilon}}\left(n^{5.246}\right)$
[JLSW20, LSW15]	CPM	$\widetilde{\mathcal{O}}_{n,R,\frac{1}{\epsilon}}\left(n^{6}\right)$
[vAG19]	QMWU	$\widetilde{\mathcal{O}}_{n,\frac{1}{\epsilon_{\mathrm{abs}}}}\left(\left(n^2 + n^{1.5}\frac{Rr}{\epsilon_{\mathrm{abs}}}\right)\left(\frac{Rr}{\epsilon_{\mathrm{abs}}}\right)^4\right)$
This work (Quantum)	IF-QIPM	$\widetilde{\mathcal{O}}_{n,\kappa,\frac{1}{\epsilon}}\left(\sqrt{n}\left(n^2\kappa_U^2 + n^4\right)\right)$
This work (Classical)	IF-IPM	$\widetilde{\mathcal{O}}_{\kappa,\frac{1}{\epsilon}}\left(n^{4.5}\kappa_U\right)$

Table 5.2: Total running times for classical and quantum algorithms to solve (5.1)-(5.2) with $m = \mathcal{O}(n^2)$.

When compared to the QMWU algorithm, it can be observed that the QIPM achieves a favorable dependence on the error. Although the QMWU is faster with respect to dimension, the QMWU has polynomial dependence on the trace bound R, and for certain applications this may negate any speedup, such as the SDO approximation of Quadratic Unconstrained Binary Optimization (QUBO) problems, i.e., problems of the form

$$\max_{x \in \{-1,1\}^n} x^\top C x. \tag{5.47}$$

QUBOs play an important role in quantum information sciences, and the resulting SDO approximation is given by

$$\max_{X \in \mathcal{S}^n} \left\{ \mathrm{tr}(CX) : \mathrm{diag}(X) = e, X \succeq 0 \right\}, \tag{5.48}$$

where e is the all ones vector of dimension n. Hence, any optimal solution of (5.48) satisfies $\mathrm{tr}(X) = n$, suggesting that for these instances the worst case running time of the QMWU is

$$\widetilde{\mathcal{O}}_{n,\frac{1}{\epsilon_{\mathrm{abs}}}}\left(n^{6.5}\left(\frac{r}{\epsilon_{\mathrm{abs}}}\right)^5\right),$$

as $R = n$ and $m = n$ for (5.48). Thus, the IF-QIPM could outperform the QMWU for this class of SDOPs, provided that $\kappa = \mathcal{O}\left(nr^{2.5}\left(\frac{1}{\epsilon}\right)^2\right)$.

For SDOPs with diagonal constraints such as (5.48), the Arora-Kale method can be specialized (see, e.g., [LP20]) to obtain an $\widetilde{\mathcal{O}}(ns\epsilon^{-3.5})$ algorithm. Note that to achieve this running time, the algorithm avoids explicitly computing or reporting the primal solution X, and instead makes use of techniques to estimate its diagonal entries and trace inner products of the form $A \bullet X$. As an alternative, these methods report a "gradient" $G \in \mathcal{S}^n$ such that $X = W \exp(G) W$ for a diagonal matrix W. While the algorithm of Lee and Padmanabhan [LP20] offers speedups with respect to the dimension, its running time is exponentially slower than classical IPMs with respect to the inverse precision.

To obtain an additive ϵ-convergence using the algorithm found in [LP20], the error parameter would have to be set to

$$\tilde{\epsilon} = \frac{\epsilon}{\|C\|_{\ell_1}}.$$

When C is dense and $\|C\|_{\ell_1} = \mathcal{O}(n)$, the running time of their algorithm therefore scales as

$$\mathcal{O}(n^2(\tilde{\epsilon})^{-3.5})) = \mathcal{O}\left(n^2\|C\|_{\ell_1}^{3.5}\epsilon^{-3.5}\right).$$

For our QIPM to reach duality gap ϵ would require complexity

$$\tilde{\mathcal{O}}_{n,\kappa,\frac{1}{\epsilon}}\left(\sqrt{n}\left(n^2\kappa_U^2 + n^4\right)\right).$$

However, one could solve the QUBO problem to ϵ-optimality using the IPM from Jiang et al. [JKL$^+$20] in time $\mathcal{O}(n^{\omega+0.5}\log(n/\epsilon))$, or even our classical IF-IPM with overall running time $\tilde{\mathcal{O}}_{\kappa,\frac{n}{\epsilon}}\left(n^{4.5}\kappa\right)$.

5.7.4   Further considerations

5.7.4.1   Choosing a starting point for the refining iterates

Now in any feasible IPM scheme it is necessary to assume the existence of a strictly feasible point $\left(X^{(0)}, y^{(0)}, S^{(0)}\right)$ such that the IPC is satisfied. As we have stressed, this can be made without loss of generality because we can always use the Self-Dual Embedding model, for which there exists a trivial interior feasible point. However, for both theory and practice, it is worthwhile to show how one can easily obtain an initial interior point for each refining problem.

The following result generalizes [MFWT21, Theorem 6.4] from LO to SDO.

Theorem 5.29 (Adapted from Theorem 6.4 in [MFWT21]). *Let* $(X^{(0)}, y^{(0)}, S^{(0)}) \in \mathcal{S}_{++}^n \times \mathbb{R}^m \times \mathcal{S}_{++}^n$ *be an interior feasible solution to the original primal-dual SDO pair* (5.1)-(5.2), *and let* $\mu^{(0)}$ *be the noramlized duality gap at this solution. Let* $\left(X^{(k)}, y^{(k)}, S^{(k)}\right) \in \mathcal{S}_+^n \times \mathbb{R}^m \times \mathcal{S}_+^n$ *and* $\eta^{(k)}$ *be the solution and scaling factor at iterate k of Algorithm ??, respectively. Then, the point*

$$\left(\eta^{(k)}\left(X - X^{(k)}\right), \eta^{(k)}\left(y - y^{(k)}\right), \eta^{(k)}S^{(0)}\right) \in \mathcal{S}^n \times \mathbb{R}^m \times \mathcal{S}_{++}^n$$

is a strictly feasible solution to the refining problem at iteration $k+1$. Moreover, the normalized duality gap at this starting point is $\left(\eta^{(k)}\right)^2\mu^{(0)}$.

Proof. First, note that for all $i \in [m]$, we have

$$\mathrm{tr}\left(A_i\left[\eta^{(k)}\left(X^{(0)} - X^{(k)}\right)\right]\right) = \eta^{(k)}\left(\mathrm{tr}\left(A_iX^{(0)}\right) - \mathrm{tr}\left(A_iX^{(k)}\right)\right) = \eta^{(k)}(b_i - b_i) = 0.$$

Similarly,

$$\sum_{i\in[m]} A_i^{\mathsf{T}}\left[\eta^{(k)}\left(y-y^{(k)}\right)\right] + \eta^{(k)}S^{(0)} = \eta^{(k)}\left(\sum_{i\in[m]} A_i^{\mathsf{T}}y - \sum_{i\in[m]} A_i^{\mathsf{T}}y^{(k)} + S^{(0)}\right)$$

$$= \eta^{(k)}\left(C - \sum_{i\in[m]} A_i^{\mathsf{T}}y^{(k)}\right)$$

$$= \eta^{(k)}\overline{C}^{(k)}.$$

Now, for any strictly feasible solution $\bar{X}$ to the refining problem, we must have

$$X^{(k)} + \frac{1}{\eta^{(k)}}\bar{X}^{(k)} \succ 0,$$

which can be equivalently expressed as

$$X^{(k)} + \frac{1}{\eta^{(k)}}\bar{X} \succ 0 \iff \bar{X} \succ -\eta^{(k)}X^{(k)}.$$

Noting that $X^{(0)} \succ 0$ by definition, the point $\eta^{(k)}\left(X - X^{(k)}\right)$ is clearly strictly feasible, as

$$\eta^{(k)}\left(X^{(0)} - X^{(k)}\right) = \eta^{(k)}X^{(0)} - \eta^{(k)}X^{(k)} \succ -\eta^{(k)}X^{(k)}.$$

We must also have $\eta^{(k)}S^{(0)} \succ 0$ because $S^{(0)} \succ 0$ and $\eta^{(k)}$ is strictly positive for all k.

Therefore,

$$\left(\eta^{(k)}\left(X^{(0)} - X^{(k)}\right), \eta^{(k)}\left(y - y^{(k)}\right), \eta^{(k)}S^{(0)}\right) \in \mathcal{S}^n \times \mathbb{R}^m \times \mathcal{S}^n_{++}$$

is a strictly feasible solution to the refining problem at iteration k, and is a valid initial point for the QIPM. This completes the proof up to trivial computation. $\qquad\square$

5.7.4.2 Barrier functions for the refining problems

As we discussed in section 2.4, self-concordant barrier functions are the foundation upon which polynomial complexity results for IPMs were built. In the case of SDO, the standard choice for the barrier function $F(X) : \mathcal{S}^n_{++} \mapsto \mathbb{R}$ involves the log-determinant function:

$$F(X) := -\log\det(X).$$

Following [dK06], define $F_P(X) := F(X)$ to be the *primal barrier*, and

$$F_D(S) := \frac{1}{\mu}b^{\mathsf{T}}y - F(S)$$

the *dual barrier*. In this case, the *primal-dual barrier* function takes the form

$$F_{PD} := F_P(X) - F_D(S) - n - n\log(\mu).$$

The question we address in this section is how to construct self-concordant barrier functions for the refining problems. Letting $X^{(k)}$ be the current primal solution and $\eta^{(k)}$ to be our scaling factor, the set of strictly feasible points to the primal refining problem associated with this solution will by defined as

$$\left\{ X \in \mathcal{S}^n : \text{tr}(A_i X) = 0 \text{ for } i \in [m], X^{(k)} + \frac{1}{\eta^{(k)}} X \succ 0 \right\}.$$

Unless $X^{(k)}$ has minimum eigenvalue 0, the refining problem is defined over a *pertubation* of the PSD cone. While we cannot simply take the barrier to the refining problem to be $-\log\det(X)$ in this case (because X need not necessarily be positive definite), there is however a simple solution.

Note that we can simply define

$$F_{\bar{P}}\left(X; X^{(k)}\right) := -\log\det\left(X^{(k)} + \frac{1}{\eta^{(k)}} X\right).$$

Clearly, $F_{\bar{P}}$ is a valid barrier for the refining problem $(\bar{P})$ induced by $X^{(k)}$; the function $F_{\bar{P}}$ diverges to infinity at the boundary of the set of matrices

$$\left\{ X \in \mathcal{S}^n : X \succeq \eta^{(k)} X^{(k)} \right\}.$$

A straightforward re-parameterization $\tilde{X} := X^{(k)} + \frac{1}{\eta^{(k)}} X$ suffices to show that of $F_{\bar{P}}(\tilde{X})$ inherits the properties enjoyed by usual barrier function F_P for the PSD cone. Adapting $F_{\bar{D}}$ from F_D in a similar manner, it follows that a valid primal-dual barrier for the refining problems is given by

$$F_{\bar{P}\bar{D}} = F_{\bar{P}} - F_{\bar{D}} - n - n\log(\mu).$$

Chapter 6

Quantum Interior Point Methods for Second Order Conic Optimization

Second Order Conic Optimization (SOCO) lies somewhere between linear optimization and semidefinite optimization (SDO); it generalizes the former, and be casted as a special case of the latter. Following the development of IPMs for SDO, the standard approach was to reformulate the problem using SDO. However, in SOCO problems, the number of variables may be much larger than the number of cones, and accordingly, IPM algorithms that could directly optimize over the Second Order Cone were developed.

In this chapter, we adapt our Inexact-Feasible IPM framework we applied to SDO to SOCO, utilizing our Iterative Refinement (IR) method for Linear systems from Chapter 4 as an efficient quantum subroutine for solving the Newton linear system. Combining these techniques, we devise the first provably convergent quantum algorithm designed for optimization over the second order cone.

6.1 Introduction

In Second-Order Cone Optimization (SOCO) [ART03, BBV04, NN95], one seeks to minimize a linear objective function subject to a feasible region defined by the intersection of an affine space with the Cartesian product of a finite number of second-order cones. Interchangeably referred to as the quadratic, *Lorentz* or "ice-cream" cone, a second-order cone can be defined in $\mathbb{R}^n$ as follows:

$$\mathcal{K} = \left\{ (x_1, x_2, \ldots, x_n) \in \mathbb{R}^n : x_1^2 - \sum_{i=2}^n x_i^2 \geq 0, x_1 \geq 0 \right\}.$$

Second-order cones admit a matrix representation; for any $x \in \mathbb{R}^n$, consider the matrix

$$\mathrm{Arw}(x) = \begin{pmatrix} x_1 & x_{2:n} \\ x_{2:n}^\top & x_1 I_{n-1} \end{pmatrix}, \tag{6.1}$$

where $x_{2:n} \equiv (x_2, x_3, \ldots, x_n)$ and I_{n-1} denotes the identity matrix of order $n - 1$. It can be seen that $x \in \mathcal{K}$ if and only if $\mathrm{Arw}(x)$ is positive semidefinite. As a consequence, SOCO can be casted a special case of Semidefinite Optimization (SDO). Further, LO is a special case of SOCO.

Given this relationship to LO and SDO, it should come as no surprise that there are polynomial-time algorithms for SOCO [ART03, NN95]. As discussed in [NN95], one approach to solving SOCO problems (SOCOPs) is to recast SOCO as a special class of SDO and apply an IPM to the resulting SDO reformulation. However, IPMs for SOCO allow an iteration bound that depends on the number of cones r, whereas the iteration bound for the SDO approach depends on the number of variables n. In SOCO, the number of variables may be much larger than the number of second order cones, implying that the direct approach is preferred [NN95].

Significant efforts were thus made to adapt the search direction results of SDO to SOCO. An important step in this area of research was made by Alder and Alizadeh [AA95], who developed the SOCO analog of the Alizadeh-Haeberly-Overton (AHO) search direction for SDO [AHO98]. Likewise, Faybusovich provided analysis for the Nesterov-Todd method for SOCO [Fay97]. Both Monteiro [MT00] and Tsuchiya [Tsu97] analyzed various search directions for SOCO using Jordan algebraic techniques. In our work, we choose the Nesterov-Todd (NT) [NT97, TTT98] direction, though our analysis generalizes for any member of the Monteiro-Zhang family of search directions for SOCO.

SOCO has historically received a considerable amount of attention, as many optimization problems can be expressed as SOCOPs (see e.g., [JKZ98, LVBL98, XY97]). Notable applications include quadratically constrained convex quadratic optimization, robust linear optimization, portfolio optimization problems such as those presented by Markowitz [BBV04, Mar59], and the support vector machine training problem (SVM) [CV95]. In fact, portfolio optimization and SVM have each been proposed as applications for quantum computing [KPS19a, RL18, RML14], however, efforts to develop a convergent quantum algorithm for directly solving SOCOPs have been unsuccessful.

6.1.1 Prior work

The lone attempt at designing a quantum algorithm for SOCO was the Quantum Interior Point Method (QIPM) presented by Kerenidis and Prakash in [KPS19b]. At a high level, their idea is to use a quantum linear system algorithm (QLSA) to obtain a speed up for the bottleneck of the classical IPM; solving the Newton linear system. Specifically, in each iteration the Newton system coefficient matrix is block-encoded using quantum random access memory (QRAM) and subsequently solved with a QLSA developed by [CGJ19], which can be accomplished in time polylogarithmic in the dimension of the problem (though this excludes the time to extract a classical estimate of the solution). The claim in [KPS19b] is that an ϵ-approximate solution to an SOCOP can be found with their algorithm in time

$$\mathcal{O}\left(\sqrt{r}n\frac{\kappa\theta}{\delta} \cdot \operatorname{polylog}\left(n, \kappa, \frac{1}{\epsilon}\right)\right)$$

where ϵ is the optimality gap, r is the number of second order cones, n the total number of variables, κ is a universal upper bound for the condition number of the Newton systems, $\theta \leq \sqrt{n}$ is a factor coming from quantum linear system solvers and δ is the error parameter for the tomography algorithm.

Due to the normalization of quantum states, to obtain a ξ-precise estimate (in ℓ_2-norm accuracy) of the solution to the Newton linear systems using the tomography algorithm in [KPS19b], one needs to set $\delta = \xi^2/\kappa^2$, meaning the running time is more accurately written as

$$\mathcal{O}\left(\sqrt{r}n\frac{\kappa^3\theta}{\xi^2} \cdot \operatorname{polylog}\left(n, \kappa, \frac{1}{\epsilon}\right)\right).$$

Further, to obtain a solution with optimality gap less than ϵ, the precision of the tomography step will need to satisfy $\xi \leq \epsilon$. Consequently, the complexity bound simplifies to

$$\mathcal{O}\left(\sqrt{r}n\frac{\kappa^3\theta}{\epsilon^2} \cdot \operatorname{polylog}\left(n, \kappa, \frac{1}{\epsilon}\right)\right).$$

However, the algorithm presented in [KPS19b] is not convergent for SOCOPs due to reasons that we discuss below. Consequently, even with the noted clarifications, the above running time does not settle the complexity of QIPMs for SOCO.

Similar to other early QIPMs (see e.g., [CMD20, KP20]), the QIPM in [KPS19b] attempts to directly quantize a classical feasible IPM, which as we saw in the case of SDO in Chapter 5, prohibits a guarantee their method will converge when applied to SOCOPs. Chief among these being the way in which the tomography errors are handled. In their analysis, the authors in [KPS19b] consider a feasible neighborhood of the central path, which requires that the primal and dual solutions remain in the interior of the primal and dual feasible sets until the algorithm terminates. This in turn only occurs when primal and dual feasibility are satisfied *exactly*, which cannot be guaranteed in the quantum regime without modifying the Newton linear system using an orthogonal subspace (or nullspace) representation [ANTZ21]. We discuss this issue from a technical standpoint in Section 6.2.6, summarize the issue as follows: the QIPM of [KPS19b] is in actuality an *infeasible* QIPM, but uses the analysis of a feasible IPM.

One could prove convergence for their method by conducting a new analysis which considers an *infeasible* central path and its neighborhood. Infeasible IPMs were originally developed due to the inherent difficulty in obtaining an interior primal-dual feasible solution to initialize the algorithm, and solving the Newton linear system directly, though we point out that infeasible QIPMs exhibit a number of disadvantages when compared to their feasible counterpart (see e.g., [ANTZ21]). When applied to SOCO, their iteration complexity $\mathcal{O}(r^2 \log(1/\epsilon))$ is worse with respect to r, and other aspects of the framework would lead to a far less efficient algorithm compared to the QIPM we present here.

Moreover, there is no consideration for issue of symmetrizing the Newton linear system to obtain the Monteiro-Zhang search directions for SOCO and convergence results for the neighborhood they consider are only guaranteed when using these search directions. In particular, the authors in [KPS19b] consider the AHO direction, which is known to be well defined only in the $\mathcal{N}_\infty(\gamma)$ neighborhood of the central path with $\gamma \in (0, \frac{1}{3})$ (see, e.g., [MT00]); which is not the neighborhood used in the analysis of [KPS19b].

6.1.2 Our results

In this chapter, we present a short-step primal-dual IF-QIPM for SOCOPs that follows the general scheme of [KPS19b], but we take safeguards from the classical IPM literature in order to ensure the validity and convergence of the algorithm on SOCOPs. In particular, we *(i)* modify the Newton linear system using an orthogonal subspaces representation of the search directions as in [ANTZ21]; *(ii)* we describe how to construct the corresponding linear system with the use of quantum-accessible data structures, making use of the block encoding techniques from [CGJ19, GSLW19] along with the iterative refinement schemes of [MAF$^+$22, MAFT22]; *(iii)* develop a novel iterative refinement scheme for SOCOPs; and *(iv)* provide a detailed analysis of the corresponding running time.

The algorithm presented here converges to an ϵ-optimal solution to the primal-dual SOCO pair (P)-(D) using at most

$$\mathcal{O}\left(\sqrt{r}n^{1.5}\kappa^2 \cdot \text{polylog}\left(n, \kappa, \frac{1}{\epsilon}\right)\right),$$

QRAM accesses and $\mathcal{O}\left(\sqrt{r}n^2 \cdot \text{polylog}\left(n, \kappa, \frac{1}{\epsilon}\right)\right)$ arithmetic operations, where r is the number of second-order cones, n is the number of variables and κ is an upper bound on the intermediate Newton system coefficient matrices that arise over the course of the algorithm. To the best of our knowledge, this yields the first provably convergent quantum algorithm designed for optimization over the second order cone. By de-quantizing our framework, we obtain a classical IF-IPM for SOCO that has worst case running time

$$\mathcal{O}\left(\sqrt{r}n\left(\kappa^2 + n\right) \cdot \mathrm{polylog}\left(n, \kappa, \frac{1}{\epsilon}\right)\right),$$

The rest of this chapter is organized as follows. In Section 6.2 we discuss the classical interior point method, in addition to important theoretical results from the IPM literature. Section 6.3 details important technical results that will play a crucial role in establishing polynomial convergence of our algorithm. In Section 6.4 we provide a Quantum Interior Point Method for SOCOPs and theoretical analysis of the running time for this Quantum Interior Point Method.

6.2   The Classical Interior Point Method for SOCOPs

In this section, we provide a detailed overview of SOCOPs, the central path and a small neighborhood of the central path.

6.2.1   The Primal and Dual SOCOPs

We consider the standard SOCOP which has the form

$$z_P = \min_{x} \left\{ \sum_{i=1}^{r} c_i^\top x_i : \sum_{i=1}^{r} A_i x_i = b, x_i \in \mathcal{K}_i, \ i \in [r] \right\}, \tag{6.2}$$

where $x_i \in \mathbb{R}^{n_i}$, $i \in [r]$ are the variables, $b \in \mathbb{R}^m$, $A_i \in \mathbb{R}^{m \times n_i}$ and $c_i \in \mathbb{R}^{n_i}$, $i \in [r]$, are the data, and $\mathcal{K}_i$ is the second order cone of dimension n_i given by

$$\mathcal{K}_i = \left\{ x_i = (x_{i0}, x_{i1}) \in \mathbb{R} \times \mathbb{R}^{n_i - 1} : x_{i0} - \|x_{i1}\| \geq 0 \right\}, \quad \forall i \in [r].$$

It is a well-known fact that second order cones are self-dual; letting $\mathcal{K}_i^*$ denote the dual cone of $\mathcal{K}_i$, we have

$$\mathcal{K}_i^* = \{s_i \in \mathbb{R}^{n_i} : s_i^\top x_i \geq 0, \ \forall x_i \in \mathcal{K}_i\} = \mathcal{K}_i \quad \forall i \in [r].$$

The dual problem associated with the primal problem is then

$$z_D = \max_{(s,y)} \left\{ b^\top y : A_i^\top y + s_i = c_i, \ s_i \in \mathcal{K}_i, \forall i \in [r] \right\}. \tag{6.3}$$

Before proceeding further, let us define the quantities

$$n = n_1 + \cdots + n_r, \qquad\qquad \mathcal{K} = \mathcal{K}_1 \times \cdots \times \mathcal{K}_r,$$
$$A = (A_1\ A_2 \cdots A_r) \in \mathbb{R}^{m \times n}, \qquad\qquad c = (c_1, \ldots, c_r) \in \mathbb{R}^n,$$
$$x = (x_1, \ldots, x_r) \in \mathbb{R}^n, \qquad\qquad s = (s_1, \ldots, s_r) \in \mathbb{R}^n,$$

and the primal and dual feasible sets:

$$\mathcal{P} = \{x \in \mathcal{K} : Ax = b\}, \quad \mathcal{D} = \left\{(s, y) \in \mathcal{K} \times \mathbb{R}^m : A^\top y + s = c\right\}.$$

Without loss of generality we assume that $m \leq n$ and $\mathrm{rank}(A) = m$. Using these quantities, the primal and dual SOCO problems (6.2) and (6.3) can be rewritten as

$$z_P = \min_x \left\{c^\top x : Ax = b, x \in \mathcal{K}\right\}, \tag{P}$$

$$z_D = \max_{(s,y)} \left\{b^\top y : A^\top y + s = c, s \in \mathcal{K}\right\}. \tag{D}$$

Continuing along this line, the sets of *interior feasible solutions* of (P) and (D) are:

$$\mathcal{P}^0 \equiv \left\{x : Ax = b, x \in \mathrm{int}\,(\mathcal{K})\right\}, \quad \mathcal{D}^0 \equiv \left\{(s, y) : A^\top y + s = c,\ s \in \mathrm{int}\,(\mathcal{K})\right\},$$

respectively, where $\mathrm{int}\,(\mathcal{K})$ denotes the interior of $\mathcal{K}$.

Without loss of generality, we assume that there exists a strictly feasible point $\left(x^{(0)}, s^{(0)}, y^{(0)}\right)$ such that $\mathcal{P}^0 \times \mathcal{D}^0 \neq \emptyset$, i.e., the Interior Point Condition (IPC) is satisfied [Wri97]. This is mild assumption; just as in the case of SDO, one can easily cast the original SOCOP as a slightly larger one via the self-dual embedding model, for which a strictly feasible solution can be obtained in a straightfoward manner. See Andersen, Roos, and Terlaky [ART03] for more details. With the IPC satisfied, it follows that optimal solutions to (P) and (D) are characterized by the system

$$x^\top s = 0, \quad Ax = b, \quad A^\top y + s = c,$$

for $(x, s, y) \in \mathcal{K} \times \mathcal{K} \times \mathbb{R}^m$.

6.2.2 Euclidean Jordan algebra, the central path and Newton directions

From here we develop the primal-dual interior point method studied in this paper using the Euclidean Jordan algebra associated with second order cones following [MT00]. The Eucildean Jordan algebra for the second order cone $\mathcal{K}_i$ is defined by a bilinear mapping $\circ : \mathbb{R}^{n_i} \times \mathbb{R}^{n_i} \to \mathbb{R}^{n_i}$:

$$x_i \circ s_i = \left(x_i^\top s_i,\ x_{i0} s_{i1} + s_{i0} x_{i1}\right), \forall x_i, s_i \in \mathbb{R}^{n_i},$$

where the unit element is given by $e_i = (1, 0, \ldots, 0)$. The natural extension of the algebra for $\mathcal{K}_i$ to $\mathcal{K} = \mathcal{K}_1 \times \ldots \times \mathcal{K}_r$ is defined as

$$x \circ s = (x_1 \circ s_1, \ldots, x_r \circ s_r), \ \forall x, s \in \mathbb{R}^n,$$

with unit element $e \equiv (e_1, \ldots, e_r)$. Further, for $x \in \mathbb{R}^n$, we let $\mathrm{mat}(x)$ denote the matrix $\mathrm{diag}(X_1, \ldots, X_r)$ where

$$X_i = \begin{pmatrix} x_{i0} & x_{i1}^\top \\ x_{i1} & x_{i0}I \end{pmatrix}, \quad i = 1, \ldots, r. \tag{6.4}$$

It follows trivially that

$$x \circ s = \mathrm{mat}(x)s = \mathrm{mat}(s)x.$$

Further, $\mathrm{mat}(x)$ is a symmetric matrix, with its smallest and largest eigenvalues defined as

$$\lambda_{\min}(\mathrm{mat}(x)) = \min\left\{ x_{i0} - \|x_{i1}\| : i \in [r] \right\} \tag{6.5a}$$

$$\lambda_{\max}(\mathrm{mat}(x)) = \max\left\{ x_{i0} + \|x_{i1}\| : i \in [r] \right\}. \tag{6.5b}$$

respectively. For more details, see Lemma 2.13 of [Tsu99]. Observe that $\mathrm{mat}(x)$ is symmetric positive semidefinite if and only if $x \in \mathcal{K}$, and symmetric positive definite if and only if $x \in \mathrm{int}\,(\mathcal{K})$.

We can define the central path of our primal-dual SOCO pair (P)-(D) as the set of solutions $(x, s, y) \in \mathcal{K} \times \mathcal{K} \times \mathbb{R}^m$ that solve

$$x \circ s = \nu e \quad Ax = b, \quad A^\top y + s = c, \tag{6.6}$$

for all $\nu > 0$. When combined with our assumption that A has full row rank, the IPC implies that (6.6) has a unique solution $(x, s, y) = (x_\nu, s_\nu, y_\nu) \in \mathrm{int}\,(\mathcal{K}) \times \mathrm{int}\,(\mathcal{K}) \times \mathbb{R}^m$. Furthermore, this solution (x_ν, s_ν, y_ν) depends on $\nu > 0$ both continuously and analytically, and as $\nu \to 0$, we obtain an optimal solution to the primal-dual SOCOP [MT00].

Applying Newton's method to system (6.6), we obtain the Newton linear system

$$S\Delta x + X\Delta s = \sigma\mu e - Xs, \quad A^\top \Delta y + \Delta s = c - s - A^\top y, \quad A\Delta x = b - Ax, \tag{6.7}$$

for $(\Delta x, \Delta s, \Delta y)$, where $X \equiv \mathrm{mat}(x)$, $S \equiv \mathrm{mat}(s)$, and $\mu \equiv \mu(x, s) = x^\top s / r$. However, the system (6.7) has no solution; we address this issue in the two sections that follow.

6.2.3 Scaling, eigenvalues and neighborhoods of the central path

From here, we follow Monteiro and Tsuchiya [MT00] in discussing a group of scaling automorphims which map the cone $\mathcal{K}$ onto itself. We then use this group to characterize the scale-invariant neighborhoods of the central path.

Let

$$\mathcal{G}_i = \left\{ \lambda \tilde{T}_i : \lambda > 0, \tilde{T}_i \in \mathbb{R}^{n_i \times n_i}, \tilde{T}_i^\top J_{n_i} \tilde{T}_i = J_{n_i}, (\tilde{T}_i)_{00} > 0 \right\}, \tag{6.8}$$

where

$$J_{n_i} \equiv \begin{pmatrix} 1 & 0 \\ 0 & -I_{n_i-1} \end{pmatrix} \in \mathbb{R}^{n_i \times n_i}.$$

It follows that $\mathcal{G}_i$ is defined as the auto-morphism group of the cone $\mathcal{K}_i$; the set of all nonsingular matrices T_i satisfying $\mathcal{K}_i = T_i(\mathcal{K}_i) \equiv \{T_i x_i : x_i \in \mathcal{K}_i\}$.[1] Likewise, choosing

$$\mathcal{G} \equiv \left\{ T = \operatorname{diag}(T_1, \ldots, T_r) : T_i \in \mathcal{G}_i, i \in [r] \right\},$$

we have that $\mathcal{G}$ is a subgroup of the auto-morphism group of the cone $\mathcal{K}$.

Each element of the Lie group $F \in \mathcal{G}_i$ has a special structure which allows one to easily construct the matrix F as well as its inverse; a fact that will be very useful later in our complexity analysis. The following result from [Tsu99] formalizes this fact.

Proposition 6.1 (Proposition 2.1 [Tsu99]). *Let F be a symmetric positive definite element of $\mathcal{G}_i$. Then, F is written as*

$$F = \begin{pmatrix} u & v^\top \\ v & I + \frac{vv^\top}{1+u} \end{pmatrix}, \quad u^2 - v^\top v = 1, \quad u > 0, \quad v \in \mathbb{R}^{n_i-1}.$$

Furthermore, the inverse of F is

$$F^{-1} = \begin{pmatrix} u & -v^\top \\ -v & I + \frac{vv^\top}{1+u} \end{pmatrix},$$

obtained by just reversing the off-diagonal part of F.

The next result from [MT00] is adapted from Proposition 2.1 of [Tsu99], and provides an explicit definition of the unique $\mathcal{G} \succ 0$ that maps e to x.

Proposition 6.2 (Proposition 1 [MT00]). *For any $x \in \operatorname{int}(\mathcal{K})$, there exists a unique symmetric matrix in $\mathcal{G}$ which maps e to x given by $T_x \equiv \operatorname{diag}(T_{x_1}, \ldots, T_{x_r})$ where for all $i \in [r]$,*

$$T_{x_i} = \begin{pmatrix} x_{i0} & x_{i1}^\top \\ x_{i1} & \beta_{x_i} I + \frac{x_{i1} x_{i1}^\top}{\beta_{x_i} + x_{i0}} \end{pmatrix}, \tag{6.9}$$

and

$$\beta_{x_i} = \sqrt{x_{i0}^2 - \|x_{i1}\|^2}. \tag{6.10}$$

Moreover, T_x is positive definite.

---

[1] A proof of this fact can be found in the appendix of [MT00].

Following [MT00, Tsu99], we define the set of $2r$ eigenvalues $\left\{ \lambda_i^j : i = 1, \ldots, r, j = 0, 1 \right\}$ corresponding to $v = (v_1, \ldots, v_r) \in \mathbb{R}^{n_1} \times \cdots \times \mathbb{R}^{n_r}$ of the Cartesian algebra to be

$$\lambda_i^0 = \lambda_i^0(v) \equiv v_{i0} - \|v_{i1}\|, \quad \lambda_i^1 = \lambda_i^1(v) \equiv v_{i0} + \|v_{i1}\|, \quad \text{for } i = 1, \ldots, r.$$

Hence, $v \in \mathcal{K}$ if and only if $\lambda_i^0 \geq 0$ for all i and similarly, $v \in \text{int}\,(\mathcal{K})$ if and only if $\lambda_i^0 > 0$. For a more detailed explanation of eigenvalues in the context of SOCOPs, we refer the reader to Section 2.4 of [Tsu99].

We now use these definitions to express the neighborhoods of the central path. For any $(x, s) \in \text{int}\,(\mathcal{K}) \times \text{int}\,(\mathcal{K})$, we have the following distance metrics $d(x, s)$:

$$d_2(x, s) \equiv \sqrt{\sum_{\substack{i=1,\ldots,r, \\ j=0,1}} \left(\lambda_i^j(w_{xs}) - \mu \right)^2} = \sqrt{2}\|w_{xs} - \mu e\|,$$

$$d_\infty(x, s) \equiv \max_{\substack{i=1,\ldots,r, \\ j=0,1}} \left| \lambda_i^j(w_{xs}) - \mu \right| = \max_{i=1,\ldots,r} \left\{ \left| w_{i0} + \|w_{i1}\| - \mu \right|, \left| w_{i0} - \|w_{i1}\| - \mu \right| \right\},$$

where $\mu = \mu(x, s)$ and

$$w_{xs} = (w_1, \ldots, w_r) \equiv T_x s. \tag{6.11}$$

Then, for $\gamma \in (0, 1)$, the short-step (or Euclidean), and infinity neighborhoods of the central path are respectively given by

$$\mathcal{N}_2(\gamma) = \left\{ (x, s, y) \in \mathcal{P}^0 \times \mathcal{D}^0 : d_2(x, s) \leq \gamma\mu(x, s) \right\},$$
$$\mathcal{N}_\infty(\gamma) = \left\{ (x, s, y) \in \mathcal{P}^0 \times \mathcal{D}^0 : d_\infty(x, s) \leq \gamma\mu(x, s) \right\}.$$

One can observe that $d_\infty(x, s) \leq d_2(x, s)$ for all $(x, s) \in \text{int}\,(\mathcal{K}) \times \text{int}\,(\mathcal{K})$, and therefore $\mathcal{N}_2(\gamma) \subset \mathcal{N}_\infty(\gamma)$ [MT00]. We point out that the product $w_{xs} = T_x s$ is the SOCO analog of the quantity $X^{1/2} S X^{1/2}$ that arises in the context of SDO. This quantity constitutes the scaled dual variable when the primal variable X is scaled to the identity matrix I. For our purposes, $w_{xs} = T_x s$ is the scaled dual variable when we scale x to e, which is why we replace the distance measure $\|X^{1/2} S X^{1/2} - \mu I\|_F$ that arises in the context of SDO with $\sqrt{2}\|w_{xs} - \mu e\|$.

The following result from [Tsu99] establishes the invariance properties of the eigenvalues of w_{xs}.

Proposition 6.3 (Proposition 2.4 in [Tsu99]). *Suppose that $(x, s) \in \text{int}\,(\mathcal{K}) \times \text{int}\,(\mathcal{K})$ and $G \in \mathcal{G}$. Let $(\tilde{x}, \tilde{s}) \equiv (G^\top x, G^{-1} s)$, $w \equiv w_{xs}$ and $\tilde{w} = w_{\overline{xs}}$. Then:*

(a) $\tilde{w}_{i0} = w_{i0}$ and $\|\tilde{w}_{i1}\| = \|w_{i1}\|$ for every $i = 1, \ldots, r$;

(b) $\lambda_i^j(\tilde{w}) = \lambda_i^j(w)$ for every $i = 1, \ldots, r$ and $j = 0, 1$;

(c) $d_2(\tilde{x}, \tilde{s}) = d_2(x, s)$ and $d_\infty(\tilde{x}, \tilde{s}) = d_\infty(x, s)$.

6.2.4 The MZ family of directions

In this section we discuss the MZ family directions for SOCOPs. Given that these directions arise as a natural extension of the Monteiro and Zhang family of directions for SDOPs, this family also emerges from computing the AHO direction (6.19) associated with a scaled problem, and subsequently mapping the resulting direction to the original (unscaled) space.

In particular, for a given matrix $G \in \mathcal{G}$, we consider the change of variables

$$\tilde{x} \equiv G^\top x, \quad (\tilde{s}, \tilde{y}) \equiv \left(G^{-1} s, y \right). \tag{6.12}$$

Upon scaling the data:

$$\tilde{c} \equiv G^{-1} c, \quad \widetilde{A} \equiv A G^{-\top}, \quad \widetilde{b} = b,$$

one can observe that the primal and dual SOCOPs (P) and (D) can be rewritten according to this scaling as:

$$(\widetilde{P}) \min \left\{ \tilde{c}^\top \tilde{x} : \widetilde{A}\tilde{x} = \tilde{b}, \ \tilde{x} \in \mathcal{K} \right\}, \tag{6.13}$$

$$(\widetilde{D}) \max \left\{ \tilde{b}^\top \tilde{y} : \widetilde{A}^\top \tilde{y} + \tilde{s} = \tilde{c}, \ \tilde{s} \in \mathcal{K} \right\}. \tag{6.14}$$

Now, one can observe that by Proposition 6.3, along with the relation $\mu(x, s) = \mu(\tilde{x}, \tilde{s})$, it follows

$$(\tilde{x}, \tilde{s}, \tilde{y}) \in \widetilde{\mathcal{N}}_2(\gamma) \iff (x, s, y) \in \mathcal{N}_2(\gamma), \quad (\tilde{x}, \tilde{s}, \tilde{y}) \in \widetilde{\mathcal{N}}_\infty(\gamma) \iff (x, s, y) \in \mathcal{N}_\infty(\gamma),$$

where $\widetilde{\mathcal{N}}_2$ and $\widetilde{\mathcal{N}}_\infty$ denote the 2-norm and ∞-norm neighborhoods of the central path for the scaled primal-dual SOCO pair (6.13)-(6.14). Further, for $\nu > 0$ denote a point on the central path of (6.13)-(6.14) by $(\tilde{x}_\nu, \tilde{s}_\nu, \tilde{y}_\nu)$ where $(\tilde{x}_\nu, \tilde{s}_\nu, \tilde{y}_\nu) = (G x_\nu, G^{-1} s_\nu, y_\nu)$.

Now to see how $G \in \mathcal{G}$ induces scaled Newton direction, note that a strictly feasible point $(x, s, y) \in \mathcal{P}^0 \times \mathcal{D}^0$ maps to an interior feasible point $(\tilde{x}, \tilde{s}, \tilde{y})$ as given by (6.12). At this scaled point, $(\tilde{x}, \tilde{s}, \tilde{y})$ we computed the corresponding AHO direction $(\widetilde{\Delta x}, \widetilde{\Delta s}, \widetilde{\Delta y})$, which we can map back to the original space; providing either the AHO direction, or another direction from the MZ family with scaling G. Hence,

$$(\Delta x, \Delta s, \Delta y) = (\Delta x_G, \Delta s_G, \Delta y_G)$$

where

$$\Delta x = G^{-\top} \widetilde{\Delta x}, \quad (\Delta s, \Delta y) = (G \widetilde{\Delta s}, \widetilde{\Delta y}).$$

That is; $(\Delta x, \Delta s, \Delta y) = (\Delta x_G, \Delta s_G, \Delta y_G)$ solves the system

$$\widetilde{S} G^\top \Delta x + \widetilde{X} G^{-1} \Delta s = \sigma \mu e - \widetilde{X}\tilde{s} + \varepsilon, \quad A^\top \Delta y + \Delta s = 0, \quad A \Delta x = 0, \tag{6.15}$$

with $\widetilde{X} = \mathrm{mat}(\tilde{x})$ and $\widetilde{S} = \mathrm{mat}(\tilde{s})$.

Now, the choice of G determines the scaled AHO direction $(\Delta x_G, \Delta s_G, \Delta y_G)$ at the point $(\Delta x, \Delta s, \Delta y)$. As such, there are many possible choices of G, each endowed with its own properties. For example, setting $G = I$ yields the AHO direction, whereas $G = T_s$ yields the HKM direction, and $G = T_x^{-1}$ for the dual counterpart of the HKM direction. Further, the MZ family of directions includes the Nesterov-Todd (NT) direction $G = G_{xs}$, where G_{xs} is the unique positive semidefinite matrix satisfying $G^2 x = s$.

Therefore, all of the results presented in this section for the unscaled AHO direction apply to every direction in the MZ family of directions, as we have demonstrated that every such direction reduces to the AHO direction in the scaled space, along with the fact that the centrality measures and duality gap are scale invariant. These results are summarized in the following result from [MT00] and one additional result for the quantum Newton linear system.

Corollary 6.1 (Corollary 2 in [MT00]). *Let $G \in \mathcal{G}$ and $(x, s, y) \in \mathrm{int}\,(\mathcal{K}) \times \mathrm{int}\,(\mathcal{K}) \times \mathbb{R}^m$ be a point such that*

$$\max_{i,j} \left| \lambda_i^j(w_{xs}) - v \right| \le \gamma\nu,$$

for some scalars $\gamma \in (0, 1/3)$ and $\nu > 0$. Then, the system (6.15) has a unique solution.

Corollary 6.2. *If the system (6.15) has a unique solution, then the quantum Newton linear system (6.17) has a unique solution.*

Proof. The proof follows exactly from the proof of Theorem 6.10. $\qquad\square$

Next, we will analyze the computational cost of classically preparing G, and its inverse G^{-1} for the Nesterov-Todd and HKM directions.

6.2.4.1 Classical cost of scaling the problem

Following the discussion found in Section 3.2 of [Tsu99], consider the matrix

$$H_i(x_i, s_i) = \omega \begin{pmatrix} u & v^\top \\ v & I + \frac{vv^\top}{1+u} \end{pmatrix}$$

where

$$u = \frac{\tau_0}{\beta_\tau}, \quad v = \frac{\tau_1}{\beta_\tau},$$

$$\tau = (\tau_0, \tau_1) = (\bar{s}_0 + \bar{x}_0, \bar{s}_1 - \bar{x}_1),$$

$$\bar{s} = (\bar{s}_0, \bar{s}_1) = \omega^{-1}(s_{i0}, s_{i1}), \quad \bar{x} = (\bar{x}_0, \bar{x}_1) = \omega(x_{i0}, x_{i1}),$$

$$\omega = \left[\frac{s_{i0}^2 - \|s_{i1}\|^2}{x_{i0}^2 - \|x_{i1}\|^2} \right]^{1/4},$$

for $i \in [r]$ and β_τ is defined according to equation (6.10). It follows that $H_i \in \mathcal{G}_i$ is the unique positive definite matrix[2] satisfying

$$H_i^{-1} s = H_i^\top x.$$

If we scale the problem by $H = \mathrm{diag}(H_1, \ldots, H_r)$, then we guarantee that the primal and dual variables scale to the same point, and in particular setting $G = H$ we obtain the Nesterov-Todd search direction. Now, it is easy to see that given x and s, each H_i can be constructed using $\mathcal{O}(n_i)$ arithmetic operations for $i \in [r]$, and thus we can compute the Nesterov-Todd scaling matrix using $\mathcal{O}(\sum_{i \in [r]} n_i) = \mathcal{O}(n)$ arithmetic operations. Moreover, by Proposition 6.1, the inverse of the Nesterov-Todd scaling matrix can be computed with the same complexity, as the only difference between the two matrices is the sign of the off-diagonal elements. From the above, it is easy to discern that the HKM direction $G = T_s$ and its inverse can also be (classically) prepared in time $\mathcal{O}(n)$.

Next, letting $G = H$ and $G^{-1} = H^{-1}$, it can be easily checked (see, e.g., Section 3.3. in [Tsu99]) that we can transform each block column A_i of A to $\tilde{A}_i$ using $\mathcal{O}(n_i m)$ arithmetic operations for $i \in [r]$. Hence, hence transforming A to $\tilde{A}$

$$\mathcal{O}\left(\sum_{i \in [r]} n_i m\right) = \mathcal{O}(nm) = \mathcal{O}(n^2)$$

arithmetic operations, as $m \leq n$ by assumption. Likewise, using G and G^{-1} we can transform c, x and s to $\tilde{c}$, $\tilde{x}$ and $\tilde{s}$ using $\mathcal{O}\left(\sum_{i \in [r]} n_i\right) = \mathcal{O}(n)$ arithmetic operations. We therefore conclude that given x and s, the classical cost of constructing the scaling matrix G (and its inverse), and subsequently scaling Newton system data is $\mathcal{O}(n^2)$ in the worst case.

6.2.5 The prototypical algorithm

We outline a classical IPM for SOCOPs in Algorithm 6. Though we drop the notation $(\tilde{x}, \tilde{s})$ to ease the presentation, the reader should understand that this discussion is applied to the scaled problem.

The algorithm begins with an initial solution pair $(x^{(0)}, s^{(0)}, y^{(0)})$ which is strictly feasible for the SOCO primal and dual problems (P) and (D), i.e., $(x^{(0)}, s^{(0)}, y^{(0)}) \in \mathcal{P}^0 \times \mathcal{D}^0$. Moreover, this initial solution has a duality gap of

$$x^{(0)^\top} s^{(0)} = \mu^{(0)} r,$$

and a distance to the central path of

$$d(x, s) \leq \gamma \mu^{(0)}$$

<hr>

[2] For a proof of this fact, and other properties of the NT scaling matrix for SOCOPs, we refer the reader to Section 7.3 in [Tsu99]

for some $\gamma \leq (0, 1)$.

In each iteration of the classical (exact-feasible) IPM for SOCOPs, we solve the scaled Newton linear system (6.15) in order to compute the Newton direction $(\Delta x, \Delta s, \Delta y)$. We then update the solutions for the subsequent iterate using the rule:

$$\left(x^{(k+1)}, y^{(k+1)}, s^{(k+1)} \right) = \left(x^{(k)} + \Delta x, s^{(k)} + \Delta s, y^{(k)} + \Delta y \right)$$

and update $\mu = \frac{x^{(k+1)\top} s^{(k+1)}}{r}$. Once μ has been reduced to or below a desired optimality tolerance $\epsilon > 0$, we can conclude that the central path parameter has been sufficiently reduced, at which point Algorithm 6 terminates and reports the optimal solution (x^*, s^*, y^*).

Algorithm 6 Classical interior point method for SOCO problems

---

Choose constants $\gamma \in (0, 1/3)$, $\epsilon \in (0, 1)$ and $\delta \in (0, 1)$;
Set $\sigma = 1 - \delta/\sqrt{2r}$;
Choose initial point $(x^{(0)}, s^{(0)}, y^{(0)}) \in \mathcal{N}_2(\gamma)$
Set $\mu^{(0)} \leftarrow \frac{x^{(0)\top} s^{(0)}}{r}$, $k \leftarrow 0$
while $\mu > \epsilon$:

1. Solve the Newton linear system (6.15) to obtain $(\Delta x, \Delta y, \Delta s)$

2. Update solution

$$x^{(k+1)} \leftarrow x^{(k)} + \Delta x, \quad s^{(k+1)} \leftarrow s^{(k)} + \Delta s, \quad y^{(k+1)} \leftarrow y^{(k)} + \Delta y$$

3. $\mu^{(k+1)} = \frac{x^{(k+1)\top} s^{(k+1)}}{r}$

6.2.6 The Nullspace Representation of the Newton Linear System

In [KPS19b], it is assumed that the Newton linear system

$$\tilde{S} G^\top \Delta x + \tilde{X} G^{-1} \Delta s = \sigma \mu e - \tilde{X}\tilde{s} + \varepsilon, \quad A^\top \Delta y + \Delta s = 0, \quad A\Delta x = 0,$$

can be solved exactly using a quantum computer for $(\Delta x, \Delta s, \Delta y)$. Applying quantum state tomography to extract a classical estimate of the solution obtained via QLSA introduces noise to the classical representation, which will only satisfy

$$\tilde{S} G^\top \Delta x + \tilde{X} G^{-1} \Delta s = \sigma \mu e - \tilde{X}\tilde{s} + \xi_c, \quad A^\top \Delta y + \Delta s = \xi_d, \quad A\Delta x = \xi_p,$$

where (ξ_c, ξ_d, ξ_p) are the quantum errors. Therefore there is no guarantee that the classical estimates obtained over the course of the algorithm satisfy primal or dual feasibility. This of course invalidates the primal-dual IPM framework they employ. Rather, we will build on the techniques introduced in Chapter 5 to construct a system that produces search directions that exhibit some error in satisfying the complementary equation but satisfy primal and dual feasibility exactly:

As we saw in the previous chapter, the constraints stipulating primal and dual feasibility have a useful interpretation which can be leveraged to overcome the challenges that arise in quantizing QIPMs on noisy quantum devices. For primal-dual feasible solution we have $x \in \mathrm{Null}(A)$ and $s \in \mathcal{R}(A)$, where $\mathrm{Null}(A)$ and $\mathcal{R}(A)$ denote the nullspace and rowspace of $A \in \mathbb{R}^{m \times n}$, respectively. Therefore, following [ANTZ21], if we can obtain bases $\mathcal{B}_N$ and $\mathcal{B}_\mathcal{R}$ of $\mathrm{Null}(A)$ and $\mathcal{R}(A)$, respectively, we can redefine the primal and dual search directions Δx and Δs as linear combinations of the basis elements.

Due to our assumption that A has full row rank, we can choose $\mathcal{B}_\mathcal{R} = [A_1^\top, \ldots, A_r^\top]$ as our basis for the rowspace of A. As described in Chapter 5, one can obtain a basis $\mathcal{B}_N$ of $\mathrm{Null}(A)$ in a variety of ways. Having obtained $\mathcal{B}_\mathcal{R}$ and $\mathcal{B}_N$, we set

$$\Delta s = \mathcal{B}_\mathcal{R} \Delta y = -A^\top \Delta y, \tag{6.16a}$$

$$\Delta x = \mathcal{B}_N \Delta z. \tag{6.16b}$$

Hence, if we are using the AHO direction, substituting these expressions for Δx and Δs into the complementarity equation, at each iteration of our IF-QIPM for SOCOP, we can simply solve:

$$\begin{bmatrix} \widetilde{S} G^\top \mathcal{B}_N & -\widetilde{X} G^{-1} A \end{bmatrix} \begin{bmatrix} \Delta z \\ \Delta y \end{bmatrix} = \sigma \mu e - \widetilde{X} \widetilde{s} + \varepsilon \tag{6.17}$$

The next result formalizes how solving (6.17) guarantees that primal and dual feasibility to be satisfied.

Proposition 6.4. *Let $(x, s, y) \in \mathcal{P}^0 \times \mathcal{D}^0$ be given, and let $(\overline{\Delta z}, \overline{\Delta y})$ be a classical estimate of the solution to (6.17) where $(\overline{\Delta z}, \overline{\Delta y}) = (\Delta z + \xi_z, \Delta y + \xi_y)$, and ξ_z and ξ_y are the error terms of the system (6.17). Letting $\overline{\Delta x}$ and $\overline{\Delta s}$ be given by (6.16a) and (6.16b), for any stepsize α we have*

$$A(x + \alpha \overline{\Delta x}) = b, \ \text{and} \ A^\top (y + \alpha \overline{\Delta y}) + (s + \alpha \overline{\Delta s}) = c.$$

Proof. For any $(x, s, y) \in \mathcal{P}^0 \times \mathcal{D}^0$, $Ax = b$ and $A^\top y + s = c$ hold. Therefore,

$$A(x + \alpha \overline{\Delta x}) = Ax + \alpha A \overline{\Delta x} = b + \alpha A \mathcal{B}_N \Delta z = b,$$

and similarly,

$$A^\top (y + \alpha \overline{\Delta y}) + (s + \alpha \overline{\Delta s}) = \left[A^\top y + s \right] + \alpha \left[A^\top \overline{\Delta y} + \overline{\Delta s} \right] = \left[A^\top y + s \right] + \alpha \left[A^\top \overline{\Delta y} + (-\mathcal{B}_\mathcal{R} \overline{\Delta y}) \right]$$
$$= c + \alpha \left[A^\top \overline{\Delta y} + (-A^\top \overline{\Delta y}) \right] = c.$$

$\square$

6.3　Technical Results

In this section, we establish results that are necessary for proving polynomial convergence of the IF-QIPM for SOCOPs. The results are adapted from [MT99] to account for the inexactness in our solution to the complementarity equation. As we saw in Section 6.2.4, each member of the Monteiro-Zhang (MZ) family of directions for SDO in the context of SOCO reduces to the AHO direction in the scaled space. In this manner, results for the AHO direction extend to each member of the MZ family of directions. Thus, we drop the cumbersome notation $\tilde{x}, \tilde{s}$, etc., to ease the presentation, though again we stress the reader should understand that this discussion is applied to the scaled problem induced by a scaling matrix G from the MZ family.

We first prove that the AHO direction in the scaled space (6.15) is well defined in the neighborhood $\mathcal{N}_{\infty}$ for $\gamma \in (0, 1/3)$, regardless of errors introduced to the complementarity of the solution by quantum tomography.

Given $(x, s) \in \operatorname{int}(\mathcal{K}) \times \operatorname{int}(\mathcal{K})$, we follow [MT00] in defining the quantities

$$X \equiv \operatorname{mat}(x),$$

$$S \equiv \operatorname{mat}(s),$$

$$R_{xs} \equiv T_x X^{-1} S T_x, \tag{6.18a}$$

$$W_{xs} \equiv \operatorname{mat}(w_{xs}). \tag{6.18b}$$

Further, note that the triple (x, s, y) is the solution which serves as the classical counterpart to the solution of our quantum Newton linear system $|z, y\rangle$. In particular, (x, s, y) solves the system

$$S\Delta x + X\Delta s = \sigma \mu e - Xs + \varepsilon \tag{6.19a}$$

$$A^{\top}\Delta y + \Delta s = 0 \tag{6.19b}$$

$$A\Delta x = 0, \tag{6.19c}$$

where ε is a residual term which captures the errors introduced by our use of quantum state tomography. We assume that ε satisfies

$$\|\varepsilon\| \le \eta \|\sigma \mu e - Xs\| \le \eta \sigma \mu / \sqrt{2} \tag{AR1}$$

for $\eta \in (0, 1)$. Note that this assumption is mild; following the discussion in Chapter 4, our use of Algorithm 2 implies that our solutions to the Newton linear systems can be made arbitrarily precise with only polylogarithmic overhead.

Lemma 6.5 (Lemma 1 in [MT00]). *For any $x \in \operatorname{int}(\mathcal{K})$, the matrices X and T_x satisfy:*

(a) $X - T_x = U_x = \text{diag}(U_{x_i} : i = 1, \ldots, r)$, where

$$U_{x_i} = \begin{bmatrix} 0 & 0 \\ 0 & (x_{i0} - \beta_{x_i})P_{x_i} \end{bmatrix}$$

and P_{x_i} is the orthogonal projection matrix onto the subspace orthogonal to x_{i1}, namely

$$P_{x_i} \equiv I - \frac{x_{i1}x_{i1}^\top}{\|x_{i1}\|^2} \tag{6.20}$$

(b) $T_x X^{-1} = X^{-1}T_x = \text{diag}(I - x_{i0}^{-1}U_{x_i} : i = 1, \ldots, r)$; as a consequence $T_x X^{-1}e = e$;

(c) X and T_x commute and $X \succeq T_x$.

The following two lemmas from [MT00] play a crucial role in the analysis of our algorithm.

Lemma 6.6 (Lemma 2 in [MT00]). *We have $R_{xs} = \text{diag}(R_i : i = 1, \ldots, r)$, where*

$$R_i = \begin{bmatrix} w_{i0} & w_{i1}^\top \\ w_{i1} & \tilde{R}_i \end{bmatrix} \tag{6.21}$$

with $(w_{i0}, w_{i1}) = T_{x_i}s_i \in \mathbb{R} \times \mathbb{R}^{n_i - 1}$ and

$$\tilde{R}_i \equiv \frac{1}{x_{i0}}\left[w_{i1}x_{i1}^\top + \beta_{x_i}^2 s_{i0}I\right] = \frac{w_{i1}x_{i1}^\top}{x_{i0}} + \left(w_{i0} - \frac{w_{i1}^\top x_{i1}}{x_{i0}}\right)I. \tag{6.22}$$

Lemma 6.7 (Lemma 3 in [MT00]). *Let $(x, s, y) \in \text{int}\,(\mathcal{K}) \times \text{int}\,(\mathcal{K}) \times \mathbb{R}^m$ be a triple such that*

$$\max_{i,j} \left|\lambda_i^j(w_{xs}) - v\right| \leq \gamma v,$$

for some scalars $\gamma > 0$ and $v > 0$. Then,

$$\|R_{xs} - W_{xs}\| \leq 2\gamma v, \tag{6.23a}$$

$$\|W_{xs} - vI\| \leq \gamma v. \tag{6.23b}$$

As a consequence,

$$\|R_{xs} - vI\| \leq 3\gamma v.$$

The next two results from [MT00] certify that the AHO direction is well defined for points in the neighborhood $\mathcal{N}_\infty(\gamma)$ with $\gamma \in (0, 1/3)$.

Lemma 6.8 (Lemma 4 in [MT00]). *Let $(x, s, y) \in \text{int}\,(\mathcal{K}) \times \text{int}\,(\mathcal{K}) \times \mathbb{R}^m$ be a triple such that*

$$\|R_{xs} - vI\| \leq \tau v, \tag{6.24}$$

for some scalars $\tau \in (0,1)$ and $\nu > 0$. Assume that $(u,v) \in \mathbb{R}^n \times \mathbb{R}^n$ and $h \in \mathbb{R}^n$ satisfy

$$Su + Xv = h, \quad u^T v \geq 0, \tag{6.25}$$

and define $\delta_u \equiv \|T_x^{-1} u\|$ and $\delta_v \equiv \|T_x v\|$. Then,

$$\delta_u \leq \frac{\|T_x X^{-1} h\|}{(1-\tau)\nu}, \quad \delta_v \leq \frac{2\|T_x X^{-1} h\|}{1-\tau}. \tag{6.26}$$

Theorem 6.9 (Theorem 1 in [MT00]). *Let $(x,s,y) \in \mathrm{int}\,(\mathcal{K}) \times \mathrm{int}\,(\mathcal{K}) \times \mathbb{R}^m$ be a point such that*

$$\max_{i,j} \left| \lambda_i^j(w_{xs}) - v \right| \leq \gamma \nu,$$

for some scalars $\gamma \in (0,1/3)$ and $\nu > 0$. Then, the system (6.19) has a unique solution. In particular, the AHO direction is well-defined at every point $(x,s,y) \in \mathrm{int}\,(\mathcal{K}) \times \mathrm{int}\,(\mathcal{K}) \times \mathbb{R}^m$ such that $d_\infty(x,s) < \mu(x,s)/3$.

We are now in a position to certify that classical estimate of a quantum state encoding the solution solves the system (6.19), and show that this solution is unique.

Theorem 6.10. *Let $(x,s,y) \in \mathrm{int}\,(\mathcal{K}) \times \mathrm{int}\,(\mathcal{K}) \times \mathbb{R}^m$ and $(\Delta z, \Delta y)$ be a classical estimate of the solution to the quantum Newton linear system (6.17) after using tomography. Then,*

$$(\Delta x, \Delta s, \Delta y) = \left(\mathcal{B}_N \Delta z, -A^\top y, \Delta y \right)$$

is a solution to the classical Newton linear system (6.19). Moreover, if the system (6.19) has a unique solution, then the quantum Newton linear system (6.17) has a unique solution.

Proof. As in [ANTZ21], the first part of the statement can be verified trivially via substitution. Uniqueness follows from a simple proof by contradiction. Suppose, in order to arrive at a contradiction, that that the system (6.17) has two distinct solutions. Then, (6.19) cannot have a unique solution, as Δx and Δs are uniquely defined as linear combinations of orthogonal basis elements. $\qquad\square$

In what follows, we make use of the following quantities

$$x(\alpha) \equiv x + \alpha \Delta x, \qquad\qquad s(\alpha) \equiv s + \alpha \Delta s,$$

$$y(\alpha) \equiv y + \alpha \Delta y, \qquad\qquad \mu(\alpha) \equiv \frac{x(\alpha)^\top s(\alpha)}{r}.$$

In the lemmas that follow, we seek to provide a bound on

$$\sqrt{2} \left\| T_x^{-1} x(\alpha) \circ T_x s(\alpha) - \mu(\alpha) e \right\|,$$

which majorizes our centrality measure $d_2\left(x(\alpha), s(\alpha)\right)$. We make one additional assumption;

that the residual term ε is chosen to satisfy the inequality

$$\|T_x X^{-1}\varepsilon\| \leq \eta \|T_x X^{-1}(\sigma\mu e - Xs)\|. \tag{AR2}$$

Lemma 6.11. *Let $(x,s,y) \in \mathcal{P}^0 \times \mathcal{D}^0$ and let $(\Delta x, \Delta s, \Delta y)$ be a solution to the system (6.19) for some $\sigma \in \mathbb{R}$. Then for every $\alpha \in \mathbb{R}$ we have:*

$$\mu(\alpha) = (1 - \alpha + \sigma\alpha)\mu + \alpha\frac{e^\top \varepsilon}{r}, \tag{6.27a}$$

$$T_x^{-1}x(\alpha) \circ T_x s(\alpha) - \mu(\alpha)e = (1-\alpha)(w_{xs} - \mu e) + \alpha(W_{xs} - R_{xs})\widetilde{\Delta x}$$
$$+ \alpha\left[T_x X^{-1}\varepsilon - \frac{e^\top \varepsilon}{r}e\right] + \alpha^2\widetilde{\Delta x} \circ \widetilde{\Delta s}, \tag{6.27b}$$

where $\mu = \mu(x,s)$ and

$$\widetilde{\Delta x} \equiv T_x^{-1}\Delta x, \qquad \widetilde{\Delta s} \equiv T_x \Delta s. \tag{6.28}$$

Proof. Since $(x,s,y) \in \mathcal{P}^0 \times \mathcal{D}^0$ by assumption, applying (6.19b) and (6.19c), one has $\Delta x^\top \Delta s = 0$. Next, multiplying (6.19a) on the left by $e^\top$ yields

$$e^\top S\Delta x + e^\top X\Delta s = s^\top \Delta x + x^\top \Delta s = \sigma r\mu - x^\top s + e^\top \varepsilon = -(1-\sigma)r\mu + e^\top \varepsilon.$$

Hence,

$$x(\alpha)^\top s(\alpha) = (x + \alpha\Delta x)^\top (s + \alpha\Delta s) = x^\top s + \alpha\left(s^\top \Delta x + x^\top \Delta s\right)$$
$$= r\mu + \alpha[-(1-\sigma)r\mu + e^\top \varepsilon] = r\mu\left[1 - \alpha(1-\sigma)\right] + \alpha e^\top \varepsilon.$$

Thus, (6.27a) follows from dividing both sides of this equality by n. On the other hand, multiplying (6.19a) on the left by $T_x X^{-1}$, as well as applying (6.11), (6.18a), (6.28) and Lemma 6.5(b), yields

$$\widetilde{\Delta s} = T_x \Delta s = T_x X^{-1}\left(\sigma\mu e - Xs + \varepsilon - S\Delta x\right) = \sigma\mu e - w_{xs} + T_x X^{-1}\varepsilon - R_{xs}\widetilde{\Delta x}.$$

We then apply this relation, along with (6.11), (6.18b), (6.28) and the fact that $u \circ v = \mathrm{mat}(u)v$ for all $u, v \in \mathbb{R}^K$, from which it follows

$$T_x^{-1}x(\alpha) \circ T_x s(\alpha) = T_x^{-1}(x + \alpha\Delta x) \circ T_x(s + \alpha\Delta s)$$
$$= \left(e + \alpha\widetilde{\Delta x}\right) \circ \left(w_{xs} + \alpha\widetilde{\Delta s}\right)$$
$$= w_{xs} + \alpha\left(w_{xs} \circ \widetilde{\Delta x} + \widetilde{\Delta s}\right) + \alpha^2\widetilde{\Delta x} \circ \widetilde{\Delta s}$$
$$= w_{xs} + \alpha\left[\sigma\mu e - w_{xs} + T_x X^{-1}\varepsilon + (W_{xs} - R_{xs})\widetilde{\Delta x}\right] + \alpha^2\widetilde{\Delta x} \circ \widetilde{\Delta s}.$$

Combining the above result with (6.27a) yields (6.27b) and the proof is complete. $\qquad\square$

In the next result, we bound norms of terms involving the residual that appear in the our analysis.

Lemma 6.12. *Let ε be a residual term satisfying the assumptions* (AR1)-(AR2). *Then,*

$$\sqrt{2}\left\|T_x X^{-1}\varepsilon\right\| \leq \eta\gamma\sigma\mu \tag{6.29a}$$

$$\sqrt{2}\left\|T_x X^{-1}\varepsilon - \frac{e^\top\varepsilon}{r}e\right\| \leq \left\|T_x X^{-1}\varepsilon\right\|. \tag{6.29b}$$

Proof. Recall from Lemma 6.5(b) that $T_x X^{-1}e = e$. By (AR2) we have:

$$\left\|T_x X^{-1}\varepsilon\right\| \leq \eta\left\|T_x X^{-1}(\sigma\mu e - Xs)\right\| = \eta\left\|(\sigma\mu e - w_{xs})\right\| \leq \eta\gamma\sigma\mu/\sqrt{2}.$$

and therefore (6.29a) holds.

For ease of notation, we define $\tilde{\varepsilon} \equiv T_x X^{-1}\varepsilon$. Then, expanding the square and using the fact that $e^\top e = r$ along with $T_x X^{-1}e = e$, we observe

$$\begin{aligned}
\left\|\tilde{\varepsilon} - \frac{e^\top\varepsilon}{r}e\right\|^2 &= \left(\tilde{\varepsilon} - \frac{e^\top\varepsilon}{r}e\right)^\top\left(\tilde{\varepsilon} - \frac{e^\top\varepsilon}{r}e\right) \\
&= \|\tilde{\varepsilon}\|^2 - \frac{2}{r}\varepsilon^\top\left(T_x X^{-1}\right)^\top\left(e^\top\varepsilon e\right) + \frac{1}{r}(e^\top\varepsilon)^2 \\
&= \|\tilde{\varepsilon}\|^2 - \frac{2}{r}\varepsilon^\top T_x X^{-1}\left(e^\top\varepsilon e\right) + \frac{1}{r}(e^\top\varepsilon)^2 \\
&= \|\tilde{\varepsilon}\|^2 - \frac{2}{r}\varepsilon^\top \underbrace{(T_x X^{-1}e)}_{=\,e}\varepsilon^\top e + \frac{1}{r}(e^\top\varepsilon)^2 \\
&= \|\tilde{\varepsilon}\|^2 - \underbrace{\frac{1}{r}(e^\top\varepsilon)^2}_{\geq\,0} \leq \|\tilde{\varepsilon}\|^2.
\end{aligned}$$

Applying the square root throughout yields (6.29b). $\square$

The following result corresponds to the inexact analogue of Lemma 6 in [MT00], and provides bounds on the norms of the scaled search directions $\|\widetilde{\Delta x}\|$ and $\|\widetilde{\Delta s}\|$, accounting for the inexactness in solving the complementarity equation.

Lemma 6.13. *Suppose that $(x, s, y) \in \mathcal{N}_2(\gamma)$ for some constant $\gamma \in (0, 1/3)$ and let $(\Delta x, \Delta s, \Delta y)$ be the unique solution of the system* (6.19), *which is a classical estimate of the solution to the quantum Newton linear system* (6.17). *Then, the directions $\widetilde{\Delta x}$ and $\widetilde{\Delta s}$ as defined in* (6.28) *satisfy:*

$$\left\|\widetilde{\Delta x}\right\| \leq \frac{(1+\eta)\Theta}{2}, \quad \left\|\widetilde{\Delta s}\right\| \leq (1+\eta)\Theta\mu, \tag{6.30}$$

where $\eta \in (0, 1)$, $\mu \equiv \mu(x, s)$ and

$$\Theta = \frac{2\left[\gamma^2/2 + (1-\sigma)^2 r\right]^{1/2}}{1 - 3\gamma}.$$

Proof. First we note that $w_{xs}^\top e = s^\top T_x e = s^\top x = r\mu$ and $\|w_{xs} - \mu e\| \le \gamma\mu/\sqrt{2}$. Hence,

$$\|w_{xs} - \sigma\mu e\|^2 = \|w_{xs} - \mu e\|^2 + \|\mu e - \sigma\mu e\|^2 + 2(1-\sigma)\mu(w_{xs} - \mu e)^\top e$$

$$\le \left[\frac{\gamma^2}{2} + (1-\sigma)^2 r\right]\mu^2. \tag{6.31}$$

Due to the fact that $d_\infty(x,s) \le d_2(x,s) \le \gamma\mu$, applying Lemma 6.7 with $\nu = \mu$ one can observe that (6.24) holds for $\tau = 3\gamma < 1$, and $\nu = \mu$.

Thus, we apply (6.11), (6.31), Lemma 6.5(b) and Lemma 6.8 with $\nu = \mu$, $(u,v) = (\Delta x, \Delta s)$, $h = \sigma\mu e - Xs + \varepsilon$ and $\tau = 3\gamma$ such that

$$\left\|\widetilde{\Delta x}\right\| \le \frac{\left\|T_x X^{-1}(\sigma\mu e - Xs + \varepsilon)\right\|\mu^{-1}}{1 - 3\gamma} \le \frac{\left(\left\|T_x X^{-1}(\sigma\mu e - Xs)\right\| + \left\|T_x X^{-1}\varepsilon\right\|\right)\mu^{-1}}{1 - 3\gamma}$$

$$\le \frac{(1+\eta)\left\|T_x X^{-1}(\sigma\mu e - Xs)\right\|\mu^{-1}}{1 - 3\gamma} = \frac{(1+\eta)\Theta}{2},$$

$$\left\|\widetilde{\Delta s}\right\| \le \frac{2\left\|T_x X^{-1}(\sigma\mu e - Xs + \varepsilon)\right\|}{1 - 3\gamma} \le \frac{2\left(\left\|T_x X^{-1}(\sigma\mu e - Xs)\right\| + \left\|T_x X^{-1}\varepsilon\right\|\right)}{1 - 3\gamma}$$

$$\le \frac{2(1+\eta)\left\|T_x X^{-1}(\sigma\mu e - Xs)\right\|}{1 - 3\gamma} = (1+\eta)\Theta\mu.$$

$\square$

Lemma 6.14 (Lemma 7 in [MT00]). *Let $u_i, v_i \in \mathbb{R}^{n_i}$ for $i = 1,\ldots,r$ and define $u \equiv (u_1,\ldots,u_r)$ and $v \equiv (v_1,\ldots,v_r)$. Then,*

$$\|u \circ v\| \le \sqrt{2}\|u\|\|v\|.$$

The next result can be viewed as the inexact analogue of Lemma 8 from [MT00], in which we establish an upper bound on a quantity which majorizes our distance metric $d_2(x(\alpha), s(\alpha))$.

Lemma 6.15. *Suppose that $(x,s,y) \in \mathcal{N}_2(\gamma)$ for some constant $\gamma \in (0, 1/3)$ and let $(\Delta x, \Delta s, \Delta y)$ be the unique solution of the system (6.19) for some $\sigma \in \mathbb{R}$. Further, $(\Delta x, \Delta s, \Delta y)$ is a classical estimate of the solution $(\Delta z, \Delta y)$ to the quantum Newton linear system (6.17). Then, for any $\alpha \in [0,1]$ we have*

$$\sqrt{2}\left\|T_x^{-1} x(\alpha) \circ T_x s(\alpha) - \mu(\alpha)e\right\|$$

$$\le \left[(1-\alpha)\gamma + \sqrt{2}\alpha\gamma(1+\eta)\Theta + \alpha^2(1+3\eta)\Theta^2 + \alpha\eta\gamma\right]\mu. \tag{6.32}$$

Proof. Noting the fact that $d_\infty(x,s) \le d_2(x,s)$, it follows that the assumption made in Lemma 6.7 holds with $\nu = \mu$. We therefore can apply (6.23a) with $\nu = \mu$, (6.27b), (6.30) and Lemma

6.14, which for all $\alpha \in [0,1]$ yields:

$$\left\| T_x^{-1} x(\alpha) \circ T_x s(\alpha) - \mu(\alpha) e \right\|$$

$$= \sqrt{2} \left\| (1-\alpha)(w_{xs} - \mu e) + \alpha(W_{xs} - R_{xs})\widetilde{\Delta x} + \alpha \left[T_x X^{-1}\varepsilon - \frac{e^\top \varepsilon}{r} e \right] \right\|$$

$$\leq (1-\alpha)\sqrt{2}\|w_{xs} - \mu e\| + \alpha\sqrt{2}\,\|R_{xs} - W_{xs}\|\,\|\widetilde{\Delta x}\| + \alpha^2\sqrt{2}\|\widetilde{\Delta x} \circ \widetilde{\Delta s}\|$$

$$\quad + \alpha\sqrt{2}\left\| T_x X^{-1}\varepsilon - \frac{e^\top \varepsilon}{r} e \right\|$$

$$\leq (1-\alpha)\sqrt{2}\|w_{xs} - \mu e\| + \alpha\sqrt{2}\,\|R_{xs} - W_{xs}\|\,\|\widetilde{\Delta x}\| + \alpha^2\sqrt{2}\|\widetilde{\Delta x} \circ \widetilde{\Delta s}\|$$

$$\quad + \alpha\sqrt{2}\left\| T_x X^{-1}\varepsilon \right\|$$

$$\leq (1-\alpha)\gamma\mu + \sqrt{2}\alpha\gamma(1+\eta)\Theta\mu + 2\alpha^2\|\widetilde{\Delta x}\|\|\widetilde{\Delta s}\| + \alpha\eta\gamma\mu$$

$$\leq (1-\alpha)\gamma\mu + \sqrt{2}\alpha\gamma(1+\eta)\Theta\mu + 2\alpha^2\left[\frac{1}{2}(1+\eta)\Theta\right][(1+\eta)\Theta\mu] + \alpha\eta\gamma\mu$$

$$= (1-\alpha)\gamma\mu + \sqrt{2}\alpha\gamma(1+\eta)\Theta\mu + \alpha^2(1+\eta)^2\Theta^2\mu + \alpha\eta\gamma\mu$$

$$= (1-\alpha)\gamma\mu + \sqrt{2}\alpha\gamma(1+\eta)\Theta\mu + \alpha^2(1+2\eta+\eta^2)\Theta^2\mu + \alpha\eta\gamma\mu$$

$$\leq (1-\alpha)\gamma\mu + \sqrt{2}\alpha\gamma(1+\eta)\Theta\mu + \alpha^2(1+3\eta)\Theta^2\mu + \alpha\eta\gamma\mu$$

$$= \left[(1-\alpha)\gamma + \sqrt{2}\alpha\gamma(1+\eta)\Theta + \alpha^2(1+3\eta)\Theta^2 + \alpha\eta\gamma\right]\mu,$$

where the final inequality follows from the fact that $\eta^2 < \eta$ as $\eta \in (0,1)$. $\qquad\square$

We are now able to state the following lemma from [MT00], which shows that the left hand side of (6.32) majorizes $d(x(\alpha), s(\alpha))$.

Lemma 6.16 (Lemma 9 in [MT00]). *Suppose* $(x,s) \in \mathrm{int}\,(\mathcal{K}) \times \mathrm{int}\,(\mathcal{K})$ *and let* $\mu = \mu(x,s)$. *Then,*

$$d_2(x,s) \equiv \sqrt{2}\,\|w_{xs} - \mu\| = \min_{G \in \mathcal{G}} \sqrt{2}\,\|x_G \circ s_G - \mu e\|,$$

where $x_G \equiv G^\top x$ *and* $s_G \equiv G^{-1}s$ *for every* $G \in \mathcal{G}$.

Proof. The result follows directly from Lemma 2.10 in [Tsu99], and is provided in full detail in [MT00]. $\qquad\square$

The final lemma of this section serves to establish the feasibility of the sequence of iterates generated by our IF-QIPM scheme.

Lemma 6.17 (Lemma 10 in [MT00]). *Let* $(x,s) \in \mathcal{K} \times \mathcal{K}$ *be given. If* $x \circ s \in \mathrm{int}\,(\mathcal{K})$, *then* $(x,s) \in \mathrm{int}\,(\mathcal{K}) \times \mathrm{int}\,(\mathcal{K})$. *In particular, if* $\sqrt{2}\|x \circ s - \nu e\| \leq \gamma\nu$ *for some* $\gamma \in (0,1)$ *and* $\nu > 0$, *then* $(x,s) \in \mathrm{int}\,(\mathcal{K}) \times \mathrm{int}\,(\mathcal{K})$.

6.4 Analysis of the Inexact-Feasible Quantum Interior Point Method for SOCOPs

In this section, we present our Inexact-Feasible Quantum Interior Point Method for SOCOPs. We establish our algorithm's polynomial convergence, as well as results on implementing block encodings of the quantum Newton linear system, concluding with the overall running time of the SOCO-IF-QIPM.

6.4.1 A quantum interior point method for SOCOPs

We quantize an IF-IPM for SOCO which is presented in Algorithm 7. To initialize the algorithm, we calculate the bases $\mathcal{B}_N$ and $\mathcal{B}_\mathcal{R}$ for the nullspace and rowspace of A, and store these matrices in QRAM. We can ignore the cost of this step as these operations need to be only carried out once, and as noted in Section 6.2.6, one could equivalently recast the problem via the Self-Dual embedding model and trivially construct $\mathcal{B}_N$ and $\mathcal{B}_\mathcal{R}$ via the coefficient matrix. Moreover, transforming the problem into its nullspace form is akin to the preprocessing steps (e.g., Cholesky factorization, ordering of the basis) that take place during the initialization phase of many optimization algorithms, both in theory and practice.

Each iteration of our IF-QIPM proceeds as follows. In step (2), we classically construct the scaling matrix G associated with our choice of search direction, and its inverse G^{-1}. For our work, we choose the Nesterov-Todd scaling matrix, as it gives the strongest performance guarantees, and can be constructed in time that is negligible compared to other operations in each iteration; classically forming G and G^{-1} from (x, s) only requires time $\mathcal{O}(n)$. In step 3, we scale the data (A, b, c) to $(\tilde{A}, \tilde{b}, \tilde{c})$ and the current solution (x, s, y) to $(\tilde{x}, \tilde{s}, \tilde{y})$. Following the discussion found in Section 6.2.4.1, we can scale the data using $\mathcal{O}(n^2)$ arithmetic operations.

In step (6.16) we construct and solve the scaled quantum Newton linear system (6.17) via block encodings, and extract a classical estimate of the resulting quantum state using Algorithm 2. A detailed analysis for step (6.16) can be found in Chapter 4, which can be accomplished in time $\tilde{\mathcal{O}}_{n,\kappa}\left(n\theta\kappa\varrho + n^2\right)$, where θ is the subnormalization factor for the Newton system coefficient matrix, and ϱ is an upper bound on the norm of the solution to the Newton linear system.

From here, we map our classical estimate $(\Delta z, \Delta y)$ of the solution to the quantum Newton linear system to $(\Delta x, \Delta s, \Delta y)$, which amounts to computing the matrix vector products $\Delta s = -A^\top \Delta y$ and $\Delta x = \mathcal{B}_N \Delta z$. Finally, we classically update the current solutions to the SOCOP x, s, and y, and the central path parameter μ.

6.4.2 Polynomial convergence

In order to have a convergent algorithm, we need to ensure that the errors ξ_k introduced from QLSA are properly chosen. Yet, by virtue of our use of the iterative refinement scheme in Algorithm 2, we can always make our precision arbitrarily accurate while incurring only poly-

---

Algorithm 7 Short-step Quantum interior point method for SOCOPs

---

Input: Choose constants $\gamma \in (0, 1/3)$ and $\delta \in (0,1)$
Set $\sigma = 1 - \delta/\sqrt{2r}$.
Let $\epsilon \in (0,1)$ and $(x^{(0)}, y^{(0)}, s^{(0)}) \in \mathcal{N}_2(\gamma)$
Set $\mu_0 \leftarrow \mu(x^{(0)}, s^{(0)})$, $k \leftarrow 0$.
Compute bases $\mathcal{B}_N$ and $\mathcal{B}_{\mathcal{R}}$ of Null(A) and $\mathcal{R}(A)$, and store these matrices in QRAM.
while $\mu > \epsilon$:

1. $\mu = \frac{x^{(k)\top} s^{(k)}}{r}$

2. Classically construct the Nesterov-Todd scaling matrix G, and its inverse G^{-1}, store these matrices in QRAM

3. Scale (A, b, c) to $(\tilde{A}, \tilde{b}, \tilde{c}) = (AG, b, G^{-1}c)$, and $(x^{(k)}, s^{(k)}, y^{(k)})$ to

$$(\tilde{x}^{(k)}, \tilde{s}^{(k)}, \tilde{y}^{(k)}) = (Gx^{(k)}, G^{-1}s^{(k)}, y^{(k)}),$$

   store in QRAM

4. Use $(\tilde{x}^{(k)}, \tilde{s}^{(k)})$ to construct $(\widetilde{X}, \widetilde{S})$, store in QRAM

5. Solve the quantum Newton linear system (6.17) and extract estimate of solution using Algorithm 2

6. Map classical estimate $(\Delta z, \Delta y)$ of $|\Delta z \circ \Delta y\rangle$ to $(\Delta x, \Delta s, \Delta y)$ according to (6.16)

7. Update solution

$$x^{(k+1)} \leftarrow x^{(k)} + \Delta x, \quad s^{(k+1)} \leftarrow s^{(k)} + \Delta s, \quad y^{(k+1)} \leftarrow y^{(k)} + \Delta y$$
$$k \leftarrow k + 1$$

---

logarithmic overhead.

In the following result we procure the analysis of one iteration of the IF-QIPM for SOCOPs for chosen constants γ, δ and η.

Theorem 6.18. *Let $\gamma \in (0, 1/3)$, $\delta \in (0,1)$ and $\eta \in (0,1)$ be constants satisfying*

$$(2 + 4\eta)\frac{2(\gamma^2 + \delta^2)}{(1 - 3\gamma)^2} + \eta\gamma \leq \left[\left(1 - \frac{\delta}{2r}\right) - \frac{\sqrt{2}}{\sqrt{r}}\eta\right]\gamma. \tag{6.33}$$

Suppose that $(x, s, y) \in \mathcal{N}_2(\gamma)$ and let $(\Delta x, \Delta s, \Delta y)$ denote the solution to the system (6.19) obtained from a classical estimate of the solution to our quantum Newton linear system (6.17) with $\sigma \equiv 1 - \delta/\sqrt{2r}$. Then,

(a) $(\hat{x}, \hat{s}, \hat{y}) \equiv (x + \Delta x, s + \Delta s, y + \Delta y) \in \mathcal{N}_2(\gamma)$;

(b) $\mu(\hat{x}, \hat{s}) = (1 - \delta/\sqrt{2r})\mu(x, s)$.

Proof. In the following proof, we seek to establish that

$$\left\| T_x^{-1}x(\alpha) \circ T_x s(\alpha) - \mu(\alpha)e \right\| \leq \gamma\mu(\alpha).$$

Given that we make no assumption regarding the sign of ε, we can use the fact that $|e^\top \varepsilon| \leq$

$\|e\|\cdot\|\varepsilon\| \le \sqrt{2r}\eta\gamma\nu$ such that

$$\left|\frac{e^{\top}\varepsilon}{r}\right| \le \frac{\sqrt{2n}}{r}\eta\gamma\mu = \frac{\sqrt{2}}{\sqrt{r}}\eta\gamma\mu.$$

Then, applying (6.33), Lemma 6.15 and using the fact that $\Theta \ge \sqrt{2}\gamma$, for all $\alpha \in [0,1]$ we have

$$
\begin{aligned}
\sqrt{2}\left\|T_x^{-1}x(\alpha)\circ T_x s(\alpha) - \mu(\alpha)e\right\| &\le \left[(1-\alpha)\gamma + \alpha(2+4\eta)\Theta^2 + \alpha\eta\gamma\right]\mu \\
&= \left((1-\alpha)\gamma + \alpha(2+4\eta)\frac{2(\gamma^2+\delta^2)}{(1-3\gamma)^2} + \alpha\eta\gamma\right)\mu \\
&\le \left((1-\alpha)\gamma + \left[\left(1-\frac{\delta}{2r}\right) - \frac{\sqrt{2}}{\sqrt{r}}\eta\right]\gamma\right)\mu \\
&\le \left((1-\alpha)\gamma + \left(1-\frac{\delta}{2r}\right)\gamma\right)\mu + \gamma\frac{e^{\top}\varepsilon}{r}.
\end{aligned}
$$

Therefore, from (6.27a), it follows:

$$\sqrt{2}\left\|T_x^{-1}x(\alpha)\circ T_x s(\alpha) - \mu(\alpha)e\right\| \le \gamma(1-\alpha+\alpha\sigma)\mu + \gamma\frac{e^{\top}\varepsilon}{r} = \gamma\mu(\alpha) < \mu(\alpha).$$

Now, combining the above with Lemma 6.17, one can observe that we obviously have $T_x^{-1}x(\alpha) \in \text{int}\,(\mathcal{K})$ and $T_x s(\alpha) \in \text{int}\,(\mathcal{K})$ for each $\alpha \in [0,1]$. Thus, in view of Proposition 6.2, this means that for every $\alpha \in [0,1]$, it follows that $(x(\alpha), s(\alpha)) \in \text{int}\,(\mathcal{K}) \times \text{int}\,(\mathcal{K})$. Further, we can note that

$$
\begin{aligned}
Ax &= b, & A\Delta x &= 0, \\
A^{\top}y + s &= c, & A^{\top}\Delta y + \Delta s &= 0,
\end{aligned}
$$

and thus we have $Ax(\alpha) = b$ and $A^{\top}y(\alpha) + s(\alpha) = c$. Therefore, we have shown that

$$(x(\alpha), s(\alpha), y(\alpha)) \in \mathcal{P}^0 \times \mathcal{D}^0$$

for every $\alpha \in [0,1]$.

From here, we apply Lemma 6.16 choosing $\tilde{x} = T_x^{-1}x(\alpha)$ and $\tilde{s} = T_x s(\alpha)$, we can conclude that for any $\alpha \in [0,1]$,

$$d(x(\alpha), s(\alpha)) \le \sqrt{2}\left\|T_x^{-1}x(\alpha)\circ T_x s(\alpha) - \mu(\alpha)e\right\| \le \gamma\mu(\alpha).$$

Thus, (a) follows as $(\hat{x}, \hat{s}, \hat{y}) = (x(1), s(1), y(1)) \in \mathcal{N}_2(\gamma)$. The statement (b) follows as a consequence of (6.27a). $\qquad\square$

Corollary 6.3. *Let* $\gamma \in (0, 1/3)$, $\delta \in (0,1)$ *and* $\eta \in (0,1)$ *be constants satisfying*

$$(2+4\eta)\frac{2(\gamma^2+\delta^2)}{(1-3\gamma)^2} + \eta\gamma \le \left[\left(1-\frac{\delta}{2r}\right) - \frac{\sqrt{2}}{\sqrt{r}}\eta\right]\gamma. \tag{6.34}$$

Then, each iterate $(x^{(k)}, s^{(k)}, y^{(k)})$ generated by Algorithm 7 is an element of the neighborhood $\mathcal{N}_2(\gamma)$ and satisfies $x^{(k)^\top} s^{(k)} = (1 - \delta/\sqrt{2r})^k x^{(0)^\top} s^{(0)}$. Further, Algorithm 7 terminates in at most $\mathcal{O}(\sqrt{r}\log(1/\epsilon))$ iterations.

Examples of constants that satisfy the condition in Corollary 6.3 are $\gamma = \delta = 1/50$ and $\eta = 1/3$.

6.4.3 The quantum newton linear system

We now discuss how to efficiently implement and solve the quantum Newton linear system using block encodings. Our work is left in general terms of $G \in \mathcal{G}$, such that it extends for every member of the MZ family of directions for SOCO.

In order to take full advantage of the block encoding framework for QSLA, we pre-compute the scaling matrix G, and its inverse, classically, and store these matrices in QRAM. Then, we utilize a factorization of the Newton system so that we can construct the entire system in time $\tilde{\mathcal{O}}(1)$. In Section 6.2.4.1, we showed that one can classically construct the Nesterov-Todd and HKM scaling matrices (and their inverses) in time that is linear in K, and is therefore more efficient than forming these matrices directly as block encodings. Namely, directly block encoding G^{-1} from a block encoding of G would introduce a dependence on the condition number κ_G of G.

The following two results establish the factors for the quantum Newton linear system (6.17).

Proposition 6.19. *Let*

$$M_1 = \begin{bmatrix} \tilde{S}G^\top \mathcal{B}_N & 0 \end{bmatrix}.$$

Let $\tilde{S}$, $G^\top$ and $\mathcal{B}_N$ be stored in QRAM. Then, a $(\|M_1\|_F, O(\log n), \xi_{M_1})$-block encoding of M_1 can be constructed in time $\tilde{\mathcal{O}}(1)$.

Proof. Noting that $\tilde{S}$ is stored in QRAM, using Proposition 3.3 we can construct a $(\|\tilde{S}\|_F, \mathcal{O}(\log n), \xi_{\tilde{S}})$-block encoding of $\tilde{S}$ in time $\tilde{\mathcal{O}}_{n,\frac{1}{\xi_{\tilde{S}}}}(1)$. Next, with $G^\top$ stored in QRAM, we can again apply Proposition 3.3 and construct a $(\|G\|_F, \mathcal{O}(\log n), \xi_G)$-block encoding of $G^\top$ in time $\tilde{\mathcal{O}}_{n,\frac{1}{\xi_G}}(1)$.

From here we apply Proposition 3.6 with

$$\xi_{\tilde{S}} = \frac{\xi_1}{2\|G^\top\|_F} \text{ and } \xi_G = \frac{\xi_1}{2\|\tilde{S}\|_F},$$

yielding a $(\|\tilde{S}G^\top\|_F, \mathcal{O}(\log n), \xi_1)$-block encoding of $\tilde{S}G^\top$.

Given that $\mathcal{B}_N$ is stored in QRAM, we implement a $(\|\mathcal{B}_N\|_F, \mathcal{O}(\log n), \xi_{\mathcal{B}_N})$-block encoding of $\mathcal{B}_N$ in time $\tilde{\mathcal{O}}_{n,\frac{1}{\xi_{\mathcal{B}_N}}}(1)$. We once more apply Proposition 3.6, choosing

$$\xi_2 = \frac{\xi_{M_1}}{2\|\mathcal{B}_N\|_F} \text{ and } \xi_{\mathcal{B}_N} = \frac{\xi_{M_1}}{2\|\tilde{S}G^\top\|_F},$$

which yields a $(\|M_1\|_F, O(\log n), \xi_{M_1})$-block encoding of M_1 can be constructed in time $\tilde{\mathcal{O}}_{n,\frac{1}{\xi_{M_1}}}(1)$, as desired. $\qquad\square$

Proposition 6.20. *Let*

$$M_2 = \begin{bmatrix} 0 & -\widetilde{X}G^{-1}A \end{bmatrix}.$$

Let $\widetilde{X}$, G^{-1} *and* A *be stored in QRAM. Then, a*

$$(\|M_2\|_F, O(\log n), \xi/(\|M_2\|_F \kappa^2 \log^2 \tfrac{\kappa}{\xi}))$$

block encoding of M_2 *can be constructed in time* $\widetilde{\mathcal{O}}_{n,\kappa,\frac{1}{\xi}}(1)$.

Proof. First, given that $\widetilde{X}$ is stored in QRAM, using Proposition 3.3 we can construct a

$$(\|\widetilde{X}\|_F, O(\log n), \xi_{\widetilde{X}})$$

block encoding of $\widetilde{X}$ in time $\widetilde{\mathcal{O}}_{n,\frac{1}{\xi_{\widetilde{X}}}}(1)$. Next, with G^{-1} stored in QRAM, we can again apply Proposition 3.3 and construct a

$$(\|G^{-1}\|_F, \mathcal{O}(\log n), \xi_G)$$

block encoding of G^{-1} in time $\widetilde{\mathcal{O}}_{n,\frac{1}{\xi_G}}(1)$.

From here we apply Proposition 3.6 with

$$\xi_{\widetilde{X}} = \frac{\xi_1}{2\|G^{-1}\|_F} \text{ and } \xi_G = \frac{\xi_1}{2\|\widetilde{X}\|_F},$$

yielding a $(\|\widetilde{X}G^{-1}\|_F, \mathcal{O}(\log n), \xi_1)$-block encoding of $-XG^{-1}$ in time $\widetilde{\mathcal{O}}_{n,\frac{1}{\xi_1}}(1)$.

Noting that A is stored in QRAM, we implement a $(\|A\|_F, \mathcal{O}(\log n), \xi_A)$-block encoding of A in time $\widetilde{\mathcal{O}}(1)$. We once more apply Proposition 3.6, choosing

$$\xi_2 = \frac{\xi_{M_2}}{2\|A\|_F} \text{ and } \xi_A = \frac{\xi_{M_2}}{2\|-XG^{-1}\|_F},$$

which yields a $(\|M_2\|_F, O(\log n), \xi/(\|M_2\|_F \kappa^2 \log^2 \tfrac{\kappa}{\xi}))$-block encoding of M_2 can be constructed in time $\widetilde{\mathcal{O}}_{n,\kappa,\frac{1}{\xi}}(1)$, as desired. $\qquad\square$

The above construction reflects only one possible way to block encode factors of our Newton system constraint matrix[3], but proceeding as we have minimizes the amount of classical computation needed.

Proposition 6.21. *The quantum Newton linear system matrix* (6.17) *can be written compactly as:*

$$M_G = \begin{bmatrix} \widetilde{S}G^{\top}\mathcal{B}_N & -\widetilde{X}G^{-1}A \end{bmatrix}, \tag{6.35}$$

---

[3]One alternative would be to compute the products $\widetilde{S}G^{\top}\mathcal{B}_N$ and $-\widetilde{X}G^{-1}A$ classically and store them in QRAM, the block encodings of M_1 and M_2 could be trivially implemented in accordance with Proposition 3.3.

Let $-\widetilde{X}, \widetilde{S}, G^{-1}, G^{\top}, A$ and $\mathcal{B}_N$ be stored in QRAM. Then, a

$$(\|M_G\|_F, \mathcal{O}(\log n), \xi/(\kappa^2 \log^2 \tfrac{\kappa}{\xi}))$$

block encoding of (6.35), can be constructed in time $\widetilde{\mathcal{O}}_{n,\kappa,\frac{1}{\xi}}(1)$.

Proof. Carrying out the calculations shows that (6.35) corresponds to the Newton linear system. We construct the two following block encodings:

$$M_1 = \begin{bmatrix} \widetilde{S}G^{\top}\mathcal{B}_N & 0 \end{bmatrix} \quad M_2 = \begin{bmatrix} 0 & -\widetilde{X}G^{-1}A \end{bmatrix},$$

using Prop.s 6.19-6.20. We choose the precision of this step so that we obtain

$$(\|M_1\|_F, \mathcal{O}(\log n), \xi/(\|M_1\|_F \kappa^2 \log^2 \tfrac{\kappa}{\xi}))$$

and $(\|M_2\|_F, \mathcal{O}(\log n), \xi/(\|M_2\|_F \kappa^2 \log^2 \tfrac{\kappa}{\xi}))$-block encodings, respectively, where κ refers to the condition number of (6.35), here and below. We add these two block encodings together using Proposition 3.4, obtaining a

$$(\max\{\|M_1\|_F, \|M_2\|_F\}, \mathcal{O}(\log n), \xi/(\kappa^2 \log^2 \tfrac{\kappa}{\xi}))$$

-block encoding of (6.35), in time $\widetilde{\mathcal{O}}_{n,\kappa,\frac{1}{\xi}}(1)$. Since $\max\{\|M_1\|_F, \|M_2\|_F\} = \mathcal{O}(\|M_G\|_F)$, we obtain the claimed result. $\qquad\square$

For completeness, we now use the factorization $M_G = M_1 + M_2$ to solve the Newton system.

Theorem 6.22. *There is a quantum algorithm that given*

$$|\sigma\mu e - \widetilde{X}\widetilde{s}\rangle$$

and access to QRAM data structures encoding $\widetilde{S}$, $\widetilde{X}$, G^{-1} $G^{\top}$, $\mathcal{B}_N$, outputs a state ξ-close to $|\Delta z \circ \Delta y\rangle$ in time $\widetilde{\mathcal{O}}_{n,\kappa,\frac{1}{\xi}}(\kappa)$, using the an appropriate direction from the MZ family of directions. We can also output an estimate of $\|\Delta z \circ \Delta y\|$ with relative error δ by increasing the running time by a factor $\frac{1}{\delta}$.

Proof. This is a direct consequence of Prop. 6.21 and Thm. 3.20. $\qquad\square$

6.4.4 Complexity

Each iteration of the IF-QIPM has a cost of

$$T_{iter} = T_{NT} + T_{progress},$$

where T_{NT} is the quantum gate complexity of obtaining a classical solution to the Newton linear system, and $T_{progress}$ is the number of classical arithmetic operations required to prepare the next iteration. The next result formally bounds this expression.

Proposition 6.23. *Let $\xi^{(k)}$ be the ℓ_2-norm precision to which we solve the Newton linear system in each iteration. Then, provided access to QRAM, each iteration of the IF-QIPM detailed in Algorithm 7 requires at most*

$$\mathcal{O}\left(n^{1.5}\kappa^2 \cdot \mathrm{polylog}\left(n, \kappa, \frac{1}{\xi^{(k)}}\right)\right),$$

QRAM accesses and $\mathcal{O}\left(n^2 \cdot \mathrm{polylog}\left(n, \kappa, \frac{1}{\xi^{(k)}}\right)\right)$ arithmetic operations.

Proof. In each iteration, we must prepare and solve the NT Newton linear system (6.17), and obtain a classical estimate of the quantum state encoding its solution. For this step, we treat Algorithm 2 as a quantum subroutine that returns a classical description of the solution to the Newton system, and by Theorem 4.5 this requires at most

$$\mathcal{O}\left(n\theta\kappa\varrho \cdot \mathrm{polylog}\left(n, \kappa, \frac{1}{\xi^{(k)}}\right)\right),$$

QRAM accesses and $\mathcal{O}\left(n^2 \cdot \mathrm{polylog}\left(n, \kappa, \frac{1}{\xi^{(k)}}\right)\right)$ arithmetic operations, where $\theta = \|M_G\|_F$ and ϱ is an upper bound on the norm of the solution to the Newton linear system. Following the proof of Proposition 5.22, choosing $\varrho = \mathcal{O}\left(\|M_G^{-1}\|\right)$ and noting $\theta = \|M_G\|_F \leq \sqrt{n}\|M_G\|$, this implies a total cost of

$$\mathcal{O}\left(n^{1.5}\kappa^2 \cdot \mathrm{polylog}\left(n, \kappa, \frac{1}{\xi^{(k)}}\right)\right),$$

QRAM accesses and $\mathcal{O}\left(n^2 \cdot \mathrm{polylog}\left(n, \kappa, \frac{1}{\xi^{(k)}}\right)\right)$ arithmetic operations to solve the Newton linear system at each iterate.

From here, we must compute the primal and dual search directions Δx and Δs using our classical estimates of Δy and Δz, and update the solution (x, y, s) in order to proceed to the next iteration. Additionally, we also need to compute the scaling matrix G and its inverse. Computing ΔX and ΔS amounts to carrying out classical matrix-vector multiplication, which requires $\mathcal{O}(n^2)$ arithmetic operations. Likewise, updating the solution corresponds to standard element-wise addition, and can be accomplished using $\mathcal{O}(n)$ arithmetic operations. Following the discussion of Section 6.2.4.1, it follows that given classical access to x and s, the scaling matrix G and its inverse can be constructed with $\mathcal{O}(n^2)$ arithmetic operations in the worst case. That is, we have:

$$T_{progress} = \mathcal{O}(n^2 + n) = \mathcal{O}(n^2),$$

and the proof is complete. $\qquad\square$

The next result bounds the overall running time of the IF-IR-QIPM detailed in Algorithm

7, and is an immediate consequence of the previous corollary and the iteration bound discussed earlier.

Corollary 6.4. *A quantum implementation of Algorithm 7 with access to QRAM outputs an ϵ-optimal solution (x^*, s^*, y^*) to the primal-dual SDO pair (P)-(D) using at most*

$$\mathcal{O}\left(\sqrt{r}n^{1.5}\kappa^2 \cdot \mathrm{polylog}\left(n, \kappa, \frac{1}{\epsilon}\right)\right)$$

QRAM accesses and $\mathcal{O}\left(\sqrt{r}n^2 \cdot \mathrm{polylog}\left(n, \kappa, \frac{1}{\epsilon}\right)\right)$ arithmetic operations.

Proof. The result is a direct consequence of combining Corollary 6.3 with Proposition 6.23. □

As in Chapter 5, it is worthwhile to consider the classical algorithm obtained from dequantizing our IF-QIPM using an inexact classical linear system solver along the lines we described in Section 5.6, though we omit a detailed discussion here. We do remark, however, that this serves as the time we could use inexact linear systems algorithms in a feasible IPM framework for SOCO.

Corollary 6.5. *A classical implementation of Algorithm 7 with access to QRAM outputs an ϵ-optimal solution (x^*, s^*, y^*) to the primal-dual SDO pair (P)-(D) in time*

$$\mathcal{O}\left(\sqrt{r}n^2\kappa \cdot \mathrm{polylog}\left(n, \kappa, \frac{1}{\epsilon}\right)\right).$$

Proof. As in the proof of Corollary 5.4, the result follows from a combined application of Corollary 6.3 and [Vis13, Theorem 16.6] with $d = n$. □

In view of the running time results in Corollaries 6.4 and 6.5, the IF-QIPM and its classical counterpart obtain speedups in the number of variables n over classical approaches. However, similar to our QIPM for SDO, additional safeguards are required to ensure that the dependence on κ does not negate these speedups. This is discussed next.

Part III

SDO Solving Frameworks Based on Gibbs Sampling

Chapter 7

Solving the Semidefinite Approximation of Quadratic Unconstrained Binary Optimization Problems

Quadratic Unconstrained Binary Optimization (QUBO) problems have immediate applications in dense graph and matrix problems, as well as quantum information sciences. While, these problems cannot be solved in polynomial time, it is well known that one can obtain tight approximations of their solution through semidefinite optimization, for which their exists polynomial time algorithms.

While their dependence on other parameters suggests no overall speedup over classical methodologies, SDO solvers based on Gibbs sampling techniques may provide speedups in the low-precision regime. We exploit this fact to our advantage, and present an iterative refinement scheme for the Hamiltonian Updates algorithm of Brandão et al. [BKF22], to exponentially improve the dependence of their algorithm on precision. As a result, we obtain a classical algorithm to solve the semidefinite relaxation of QUBOs in matrix multiplication time. A quantum implementation of our algorithm with access to QRAM requires at most

$$\mathcal{O}\left(n^{1.5}\cdot\mathrm{polylog}\left(n,\|C\|_F,\frac{1}{\epsilon}\right)\right)$$

accesses to the QRAM and additional quantum gates plus $\mathcal{O}(ns)$ classical arithmetic operations.

In the sparse-access input model (without QRAM), the algorithm takes

$$\tilde{\mathcal{O}}_{n,\|C\|_F,\frac{1}{\epsilon}}\left(n^{1.5}s^{0.5+o(1)}\right)$$

accesses to an oracle describing the coefficient matrix C and $\widetilde{\mathcal{O}}_{n, \|C\|_F, \frac{1}{\varepsilon}}\left(n^{2.5} s^{0.5 + o(1)}\right)$ additional gates, therefore yielding no quantum speedup (the quantum gate complexity is asymptotically larger than the classical complexity).

7.1 Introduction

We consider optimization problems of the form:

$$\max x^\top C x$$
$$\text{s.t. } x \in \{-1, 1\}^n, \tag{7.1}$$

where $C \in \mathcal{S}^n$ is the problem data. Solving (7.1) can be viewed as computing the $\infty \to 1$ norm of the coefficient matrix C. The $\infty \to 1$ norm is intrinsically related the *cut norm* of a matrix, which is fundamental to the development of efficient approximation algorithms for dense graph and matrix problems [ADLVKK03, FK99], such as finding the largest cut in a graph (MaxCut). Problems of the form given in (7.1) also play an important role in quantum information sciences; the Ising model belongs to this class of problems [PBP13], and it has proven to be the key use case for quantum heuristics such a the Quantum Approximate Optimization Algorithm (QAOA) [FGG14] and quantum annealing [FGS$^+$94].

If we replace $x \in \{-1, 1\}^n$ with $z \in \{0, 1\}^n$ in (7.1), we obtain a *quadratically unconstrained binary optimization* (QUBO) problem. A standard QUBO is of the form

$$\max z^\top C z$$
$$\text{s.t. } z \in \{0, 1\}^n, \tag{7.2}$$

and corresponds to computing the cut norm. It is well known that one can use solutions to (7.1) to compute solutions to (7.2), provided that we allow for linear terms (in both formulations). This is due to the equivalence $z = \frac{x+e}{2}$ if $z \in \{0, 1\}^n$ and $x \in \{-1, 1\}^n$, where $e \in \mathbb{R}^n$ is the all ones vector of dimension n.

Although (7.1) and (7.2) cover many applications of interest, computing optimal solutions to either is NP-Hard in general. Following the seminal work of Lovász [Lov79] and the theoretical and practical development of Interior Point Methods (IPMs) for solving semidefinite optimization (SDO) problems [Mon98, NT97, NT98, Stu99, TTT99], a prevailing approach has been to obtain approximate solutions to (7.1) and (7.2) by relaxing integrality and lifting the problem from a vector space of dimension n, to the space of $n \times n$ symmetric matrices. The quadratic form $x^\top C x$ can be equivalently expressed by $\text{tr}\,(C x x^\top)$, where $\text{tr}\,(U)$ denotes the sum of the diagonal elements (or, trace) of a matrix $U \in \mathbb{R}^{n \times n}$. To deal with the bilinear term $x x^\top$, we introduce a matrix variable $X \in \mathbb{R}^{n \times n}$, and require that X satisfies the following:

$$\text{diag}(X) = e, \quad X \succeq 0, \quad \text{rank}(X) = 1,$$

where the notation $U \succeq V$ means that the matrix $U - V$ is a positive semidefinite matrix. Under these requirements, X is guaranteed to be of the form $X = x x^\top$ for $x \in \{-1, 1\}^n$. The rank constraint, however, is not convex, and thus dropping it yields the following (convex) SDO relaxation of (7.1):

$$\max \ \operatorname{tr}(CX)$$
$$\text{s.t. } \operatorname{diag}(X) = e, \quad X \succeq 0. \tag{QUBO-SDO}$$

Although the optimal solution X^* to (QUBO-SDO) is no longer guaranteed to satisfy $X^* = x^* x^{*^\top}$ and may not be integral in general, the approximation of x^* provided by X^* is of sufficient quality to justify its use.

In fact, SDO approximations cover some of the most celebrated results in optimization, such as the 0.878-approximation guarantee of Goemans and Williamson for MaxCut [GW95] and the Lovász-ϑ number [Lov79]. SDO problems of the form (QUBO-SDO) can also be used for relaxations of MaxSat [Anj06, dKVMW00, GVHL06] and MaxBisection [FJ97], and by adding constraints on the off-diagonal elements of our solution, we can utilize (QUBO-SDO) to approximate covariance matrices that arise in financial problems [Hig02]. For a detailed review of SDO approximations of quadratic optimization problems, we refer the reader to the survey by Luo et al. [LMS$^+$10].

7.1.1   Prior work

As we discussed in Chapter 5, two major directions have thus far been proposed to solve SDO problems using quantum computers; QIPMs and QMMWUs. Up to this point, we have focused only studied the former class of algorithms, which rely on QLSAs to obtain speedups. In this chapter, we will study a special case of the QMMWU framework, which build on the observation that normalized positive semidefinite matrices can be represented using quantum states, and do not involve the solution of linear systems.

Recall that QMMWU methods quantize algorithms based on matrix exponentials and Gibbs states, namely, the Matrix Multiplicative Weights Update (MMWU) Method of Arora and Kale [AHK12], and were first proposed by Brandão and Svore [BS17] and van Apeldoorn et al. [vAGGdW20]. Although subsequent improvements [Gri19, vAG19] have reduced the running time of QMMUs to

$$\tilde{\mathcal{O}}_{n,R,\frac{1}{\epsilon}}\left(\left(\sqrt{m} + \sqrt{n}\frac{Rr}{\epsilon}\right) s \left(\frac{Rr}{\epsilon}\right)^4\right),$$

these algorithms still exhibit a non-polynomial running time, due to their polynomial dependence on the scale invariant parameter $\frac{Rr}{\epsilon}$, whereas the natural input size depends on the logarithm of this quantity.

Naturally, the question has been raised as to whether QMMU framework could be specialized for certain applications of SDO for in order to obtain unconditional quantum speedups. In [BKF22], Brandão et al. present such an algorithm, which they call *Hamiltonian Updates* (HU), for solving the SDO approximation (QUBO-SDO) of (7.1). The HU method is a primal-only algorithm closely related to the QMMWU framework, in that it leverages a Gibbs state representation of the variables and progression towards the optimal solution is made via matrix-exponentiated gradient updates. Specifically, the authors in [BKF22] are interested in solving an SDO feasibility problem that arises upon renormalizing and relaxing (QUBO-SDO):

$$
\begin{aligned}
\text{find} \quad & X \\
\text{s.t.} \quad & \operatorname{tr}\left(\frac{C}{\|C\|}X\right) \geq \gamma - \epsilon \\
& \sum_{i \in [n]} \left| \langle i|X|i \rangle - \frac{1}{n} \right| \leq \epsilon \\
& \operatorname{tr}(X) = 1, \quad X \succeq 0.
\end{aligned}
\tag{7.3}
$$

Here, γ is an upper bound on the absolute value of the optimal objective value of (QUBO-SDO) when the cost matrix C is normalized, obtained via binary search over $[-1, 1]$, and $|i\rangle$ for $i \in \{1, \ldots, n\}$ are the computational basis states. Letting H denote the Hamiltonian associated with (7.3), any solution to (7.3) can be naturally be expressed as the Gibbs state

$$
\rho = \frac{\exp(-H)}{\operatorname{tr}(\exp(-H))},
$$

as any $\log(n)$-qubit Gibbs state is an element of the set $\{X \in \mathbb{R}^{n \times n} : \operatorname{tr}(X) = 1, X \succeq 0\}$ by definition. The key observation in [BKF22] is that upon using the Gibbs state change of variables in (7.3), one can model the n constraints on the diagonal elements as single constraint which requires that the total variation distance from the distribution along the diagonal elements of a solution to (7.3) to the maximally-mixed state $n^{-1}I$ be no more than ϵ. In other words, the task of solving (7.3) reduces to finding a $\log(n)$-qubit mixed quantum state which, upon measurement in the computational basis is approximately indistinguishable from the maximally-mixed state, and whose trace inner product with the normalized cost matrix $C\|C\|^{-1}$ is at least $\gamma - \epsilon$.

Using a quantum computer, the HU method of [BKF22] solves (QUBO-SDO) to additive error $n\|C\|\epsilon$ in time

$$
\tilde{\mathcal{O}}_{n, \frac{1}{\epsilon}}\left(n^{1.5}\sqrt{s}^{1+o(1)} \epsilon^{-28+o(1)} \exp\left(1.6\sqrt{\log(\epsilon^{-1})}\right) \right).
$$

The authors in [BKF22] also provide an analysis of essentially the same algorithm when using a classical computer, and show that the classical algorithm has a complexity of

$$
\tilde{\mathcal{O}}_n\left(\min\{n^2 s, n^\omega\} \epsilon^{-12} \right).
$$

The quantum algorithm yields a speedup in n over classical algorithms, for a specific class of SDO problems. However, as we have already seen with QIPMs and QMMWU algorithms, its dependence on other parameters (in this case the inverse precision) is prohibitive unless a very low precision solution is acceptable. This raises the question as to whether the poor scaling in the inverse precision can be mitigated without incurring additional cost in n and s. We answer this question in the affirmative using iterative refinement techniques.

7.1.2 Our results

In this chapter, we develop an iterative refinement (IR) scheme for SDO approximations of QUBO problems that utilizes the HU algorithm of [BKF22] as a subroutine. We show that proceeding in this way allows one to exponentially improve the dependence on the inverse precision for both the quantum and classical algorithms presented in [BKF22].

With the proposed IR scheme, the classical algorithm solves the SDO problem (QUBO-SDO) up to absolute error $\mathcal{O}(\epsilon)$ with worst-case complexity

$$\mathcal{O}\left(\min\{n^2 s, n^\omega\} \cdot \operatorname{polylog}\left(n, \|C\|_F, \frac{1}{\epsilon}\right)\right).$$

This is a significant speedup compared to general-purpose SDO solvers, such as IPMs. This algorithm can be quantized following a similar strategy to [BKF22]. When provided access to quantum random access memory (QRAM), the quantum algorithm takes

$$\mathcal{O}\left(n^{1.5} \cdot \operatorname{polylog}\left(n, \|C\|_F, \frac{1}{\epsilon}\right)\right)$$

accesses to the QRAM and additional quantum gates (this is the standard way of describing complexity in the QRAM model of computation), plus $\mathcal{O}(ns)$ classical arithmetic operations — note that simply reading the cost matrix C takes $\mathcal{O}(ns)$ time.

Summarizing, the combination of HU with IR described in this paper provides exponential speedups over the methodology proposed in [BKF22] with respect to the precision parameter ϵ. To the best of our knowledge, our classical and quantum algorithms are the fastest known algorithms in their respective model of computation for this class of problems, and our quantum algorithm provides a genuine asymptotic speedup over known classical solution methodologies, provided that we have access to QRAM. In the sparse-access input model (without QRAM), the algorithm takes $\widetilde{\mathcal{O}}_{n, \|C\|_F, \frac{1}{\epsilon}}\left(n^{1.5} s^{0.5 + o(1)}\right)$ accesses to an oracle describing the coefficient matrix C and $\widetilde{\mathcal{O}}_{n, \|C\|_F, \frac{1}{\epsilon}}\left(n^{2.5} s^{0.5 + o(1)}\right)$ additional gates, therefore yielding no quantum speedup (the quantum gate complexity is asymptotically larger than the classical complexity).

The remainder of this chapter is organized in the following manner. Section 7.2 introduces notation, as well as the relevant input models and quantum subroutines. In Section 7.3 we introduce the Hamiltonian Updates (HU) algorithm from [BKF22], and our Iterative Refinement scheme for SDO approximations of QUBOs is presented in Section 7.4. The running time analysis is performed in Section 7.5.

7.2 Preliminaries

Later in this work, we make use of the following facts regarding Hadamard products. Later in this work, we make use of the following facts regarding Hadamard products.

Lemma 7.1 (Lemma 5.1.4 in [HJ94]). *Let E, F and G be $m \times n$ matrices. Then, the i-th diagonal entry of the matrix $(E \circ F)G^\top$ coincides with the i-th diagonal entry of the matrix $(E \circ G)F^\top$. That is,*

$$[(E \circ F)G^\top)]_{ii} = [(E \circ G)F^\top)]_{ii} \quad \forall i \in [m].$$

Lemma 7.2 (Theorem 5.3.4 in [HJ94]). *Let A and B be $n \times n$ Hermitian matrices. If $A \in \mathcal{S}^n_+$, then any eigenvalue $\lambda(A \circ B)$ of $A \circ B$ satisfies*

$$\lambda_{\min}(A) \cdot \lambda_{\min}(B) \leq \min_{i \in [n]} A_{ii} \cdot \lambda_{\min}(B) \leq \lambda(A \circ B) \leq \max_{i \in [n]} A_{ii} \cdot \lambda_{\min}(B) \leq \lambda_{\max}(A) \cdot \lambda_{\max}(B).$$

For a scalar $x \in \mathbb{R}$ define the *sign function* $\mathrm{sign}(x)$ as

$$\mathrm{sign}(x) := \begin{cases} -1 & \text{if } x < 0 \\ 0 & \text{if } x = 0 \\ 1 & \text{if } x > 0. \end{cases}$$

When $x \in \mathbb{R}^n$, $\mathrm{sign}(x) = (\mathrm{sign}(x_1), \ldots, \mathrm{sign}(x_n))^\top$.

7.2.1 Input models and subroutines

Similar to other works on quantum SDO solvers [vAGGdW20], we provide analyses that consider two input models: the *sparse-access model*, and the *quantum operator model*. We omit a discussion on the *quantum state model* [BKL$^+$19, vAG19], as the quantum operator model generalizes both the quantum state model and the sparse-access model.

For our purposes here, in the *sparse-access model*, the input matrix C is assumed to be s-row sparse for some known bound $s \in [n]$. In other words, C has at most s nonzero entries per row. The sparse-access model is closely related to the classical notion, in that we assume access to an oracle O_{sparse}, which upon being queried with input (i, j) returns the index of the j-th nonzero entry of the i-th row of C by calculating the index function:

$$\mathrm{index} : [n] \times [s] \to [n].$$

That is, for $i \in [n]$ and $j \in [s]$, O_{sparse} computes the position in place:

$$O_{\mathrm{sparse}} |i, j\rangle = |i, \mathrm{index}(i, j)\rangle.$$

Additionally, we assume access to another oracle that returns a bitstring representation of the the magnitude of individual entries for every $i, j \in [n]$:

$$O_C |i, j, z\rangle = |i, j, z \oplus (C_{ij} \|C\|_F^{-1})\rangle.$$

7.3   Hamiltonian Updates

In this section, we present the algorithm from [BKF22] and relevant results required to prove its convergence and analyze its cost.

7.3.1   Convex Feasibility Problems

In order to avoid any normalization issues for the problems that arise over the course of our IR scheme, we deviate slightly from [BKF22] and renormalize the problem (QUBO-SDO) using the Frobenius norm of the cost matrix rather than use its operator norm:

$$
\begin{aligned}
\text{find} \quad & X \\
\text{subject to} \quad & \operatorname{tr}\left(\frac{C}{\|C\|_F}X\right) \geq \gamma - \epsilon \\
& \sum_{i \in [n]} \left| \langle i|X|i\rangle - \frac{1}{n} \right| \leq \epsilon \\
& \operatorname{tr}(X) = 1, \quad X \succeq 0.
\end{aligned}
\qquad \text{(OptRelaxed)}
$$

The relaxed renormalized SDO problem (OptRelaxed) is a specific example of the convex optimization problem

$$
\begin{aligned}
\max \quad & f(X) \\
\text{subject to} \quad & X \in \mathcal{P}_1 \cap \mathcal{P}_2 \cap \cdots \cap \mathcal{P}_m, \\
& \operatorname{tr}(X) = 1, \ X \succeq 0,
\end{aligned}
\qquad (7.4)
$$

in which $f(X)$ is a convex function of X and $\mathcal{P}_1, \ldots, \mathcal{P}_m$ are convex sets.

In this context, the trace constraint enforces normalization, but also allows us to obtain a bound on the optimal objective value. Letting $\tilde{C} = C\|C\|_F^{-1}$ and invoking the tracial matrix Hölder inequality [Bha13], it follows that any X^* that solves (7.4) satisfies the following relation:

$$
\left| \operatorname{tr}(\tilde{C}X^*) \right| \leq \|\tilde{C}\| \|X^*\|_{\operatorname{tr}} = \|\tilde{C}\|.
$$

It is well known in the optimization literature that performing binary search over the range of values

$$
\gamma \in \left[-\|\tilde{C}\|, \|\tilde{C}\| \right] \subseteq [-1, 1]
$$

that the objective can take reduces the task of solving (7.4) to solving a sequence of feasibility problems of the form:

$$
\begin{aligned}
\text{find} \quad & X \in \mathcal{S}_+^n \cap \{X : \operatorname{tr}(X) = 1\} \\
\text{subject to} \quad & \operatorname{tr}(\tilde{C}X) \geq \gamma \\
& X \in \mathcal{P}_1 \cap \mathcal{P}_2 \cap \cdots \cap \mathcal{P}_m.
\end{aligned}
\qquad (7.5)
$$

In particular, $\log(\|\tilde{C}\|\epsilon^{-1}) = \mathcal{O}\left(\log(\epsilon^{-1})\right)$ queries to (7.5) are sufficient to estimate the optimal objective value of (7.4) up to additive error ϵ.

7.3.2 Solving Convex Feasibility Problems via Hamiltonian Updates

Hamiltonian Updates (HU) is a meta-algorithm for solving convex feasibility problems of the form (7.5), adapted from the work of Tsuda, Rätsch and Warmuth [TRW05] as well as [AK16, BKL$^+$19, Haz16, LSW15]. At a high level, HU can be viewed as a mirror descent algorithm [Nem79, NY83] with the negative von Neumann entropy as the mirror map.[1] In each iteration, the method uses certain subroutines to test ϵ-closeness to convex sets $\mathcal{P}_1, \mathcal{P}_2, \ldots, \mathcal{P}_m$, which we formally define next.

Definition 7.1 (Definition 2.1 in [BKF22]). *Let $\mathcal{P} \subset \{X \in \mathcal{S}_+^n : \mathrm{tr}(X) = 1\}$ be a closed, convex subset of quantum states, and $\widetilde{\mathcal{P}} \subset \{X \in \mathbb{C}^{n \times n} : X = X^\dagger, \|X\| \leq 1\}$ be a closed, convex subset of observables of operator norm at most 1. For $\epsilon > 0$, an ϵ-separation oracle with respect to $\widetilde{\mathcal{P}}$ is a subroutine that either accepts a state ρ (in the sense that observables from $\widetilde{\mathcal{P}}$ cannot distinguish ρ from the elements of $\mathcal{P}$), or provides a normal vector (in the matrix space) P of a hyperplane that separates ρ from the set $\mathcal{P}$ using a test from $\widetilde{\mathcal{P}}$:*

$$O_{\mathcal{P},\epsilon}(\rho) = \begin{cases} \textit{accept } \rho & \textit{if } \min_{Y \in \mathcal{P}} \max_{P \in \widetilde{\mathcal{P}}} \mathrm{tr}(P(\rho - Y)) \leq \epsilon, \\[2ex] \textit{output } P \in \widetilde{\mathcal{P}} \textit{ s.t. } \mathrm{tr}(P(\rho - Y)) \geq \frac{\epsilon}{2} \textit{ for all } Y \in \mathcal{P} & \textit{otherwise.} \end{cases}$$

The authors in [BKF22] point out that the above oracle construction is well defined, as we can always choose some hyperplane $P \in \widetilde{\mathcal{P}}$ such that

$$\mathrm{tr}\left(P(\rho - Y)\right) \geq \frac{\epsilon}{2},$$

holds for all $Y \in \mathcal{P}$ whenever

$$\min_{Y \in \mathcal{P}} \max_{P \in \widetilde{\mathcal{P}}} \mathrm{tr}(P(\rho - Y)) > \epsilon.$$

From Sion's min-max theorem [Sio58], it follows that

$$\max_{P \in \widetilde{\mathcal{P}}} \min_{Y \in \mathcal{P}} \mathrm{tr}(P(\rho - Y)) = \min_{Y \in \mathcal{P}} \max_{P \in \widetilde{\mathcal{P}}} \mathrm{tr}(P(\rho - Y)) > \epsilon,$$

and hence there exists a hyperplane which separates ρ from $\mathcal{P}$ by ϵ. By relaxing the requirement to $\frac{\epsilon}{2}$-separation, the algorithm is able to reconcile with the errors that result from approximating quantities computed with ρ, or estimating its entries.

The Hamiltonian Updates (HU) algorithm of Brandão et al. [BKF22] is provided in full detail in Algorithm 8. The algorithm takes as input the precision parameter ϵ, and m ϵ-separation oracles $O_{1,\epsilon}, O_{2,\epsilon}, \ldots, O_{m,\epsilon}$. In the initialization steps, the starting point is defined to be the maximally mixed state $\rho \leftarrow n^{-1}I$. This is critical to ensuring the convergence of mirror descent-based approaches such as Algorithm 8 and the works in [AK16, BKL$^+$19, Haz16, LSW15, TRW05]; initialization to the maximally mixed state ensures that the quantum relative entropy between any

---

[1] Allen-Zhu and Orecchia show how MMWU algorithms can be derived from mirror descent in [AZO17, Appendix A.2].

feasible state and the initial state is bounded by $\log(n)$ (see, e.g., Theorem 11.8 pt. 2 [NC02]), and is reduced at every iteration. Consequently, Algorithm 8 terminates in a finite number of iterations.

As noted in [BKF22], how we define $\widetilde{\mathcal{P}}$ determines the number of closeness conditions that need to be tested. By using the Gibbs state change of variables, we do not need to test if our candidate solution is trace normalized or positive semidefinite; any Gibbs state

$$\rho_H = \frac{\exp(-H)}{\operatorname{tr}(\exp(-H))}$$

is an element of the set $\{X \in \mathcal{S}_+^n : \operatorname{tr}(X) = 1\}$ by definition. Our task therefore reduces to finding a $\log(n)$-qubit mixed state ρ which is ϵ-close to the convex sets $\mathcal{P}_i$ that arise from any other constraints included in the feasibility problem. At each iteration, ϵ-closeness is tested by querying ϵ-separation oracles which are constructed using observables in $\widetilde{\mathcal{P}}_i$. If each of our oracles accepts the candidate state, the algorithm terminates and reports (ρ, H) as an ϵ-precise solution. Otherwise, upon detecting infeasibility the matrix exponent is updated to penalize the infeasible directions using the rule

$$H \leftarrow H + \frac{\epsilon}{16}P,$$

where P is a normal vector in the matrix space of a hyperplane that witnesses infeasibility.

Algorithm 8 Hamiltonian Updates for Convex Feasibility Problems

---

Input: Query access to m ϵ-separation oracles $O_{1,\epsilon}(\cdot), \ldots, O_{m,\epsilon}(\cdot)$
Initialize $\rho = n^{-1}I$ and $H = 0^{n \times n}$
for $t = 1, \ldots, T$ do
    for $i = 1, \ldots, m$ do
        if $O_{i,\epsilon}(\rho) = P$ then
            $H \leftarrow H + \frac{\epsilon}{16}P$
            $\rho \leftarrow \frac{\exp(-H)}{\operatorname{tr}(\exp(-H))}$
            break
        end
    end
    return (ρ, H) and **exit**
end

The following result establishes the iteration complexity of Algorithm 8.

Theorem 7.3 (Theorem 2.1 in [BKF22]). *Algorithm 8 requires at most $T = \lceil 64 \log(n)\epsilon^{-2} \rceil + 1$ iterations to certify that (7.5) is infeasible or output a state ρ satisfying*

$$\text{for all } 1 \leq i \leq m : \ \max_{P_i \in \bar{\mathcal{P}}_i} \min_{Y_i \in \mathcal{P}_i} \operatorname{tr}(P_i(\rho - Y_i)) \leq \epsilon.$$

Note that Theorem 7.3 applies to *any* convex feasibility problem (on density operators, i.e., trace-normalized positive semidefinite matrices) for which we have separation oracles as outlined in Definition 7.1. This is crucial for the development of an iterative refinement scheme.

There is an important distinction with respect to output across the models of computation we study. A classical implementation of Algorithm 8 outputs an explicit description of an ϵ-

precise solution ρ^* to (OptRelaxed) and its associated Hamiltonian H^*, whereas a quantum implementation reports a real valued vector $y \in \mathbb{R}^2$ along with a diagonal matrix D (with $\|D\| \leq 1$) such that $H^* = y_1 \widetilde{C} + y_2 D$. The vector $y = (y_1, y_2)^\top$ is the *state preparation pair* of ρ^*, in particular:

$$\rho^* = \frac{\exp\left(-\left(y_1 \widetilde{C} + y_2 D\right)\right)}{\mathrm{tr}\left[\exp\left(-\left(y_1 \widetilde{C} + y_2 D\right)\right)\right]},$$

and we refer to this type of output as a *state preparation pair description* of ρ. This choice of output is used in all quantum SDO solvers based on Gibbs sampling techniques (see, e.g., [BKL+19, BKF22, BS17, vAG19, vAGGdW20]), and is motivated by the fact that it is difficult to develop quantum algorithms that are substantially faster than classical algorithms if we still have to output each entry of the solution (an $n \times n$ matrix).

The Gibbs sampling approaches that we apply later exhibit a cost that depends on a norm bound for y. Observe that we initialize y to the all zeros vector of appropriate dimension, and in every iteration, at most one entry of y changes by a magnitude of $\frac{\epsilon}{16}$ (specifically, an entry y_i, where the oracle $O_{i,\epsilon}$ has detected infeasibility). As a consequence, the vector y satisfies the inequality

$$\left\| y^{(t+1)} - y^{(t)} \right\| \leq \frac{\epsilon}{16} \tag{7.6}$$

for each iteration t. In view of the iteration bound for Algorithm 8 provided in Theorem 7.3, it is easy to see that for any y obtained from Algorithm 8 we have

$$\|y\|_1 \leq \lceil 64 \log(n)\epsilon^{-2} \rceil \left\| y^{(t+1)} - y^{(t)} \right\| \leq \lceil 64 \log(n)\epsilon^{-2} \rceil \frac{\epsilon}{16} \leq 4 \log(n)\epsilon^{-1}. \tag{7.7}$$

To instantiate the algorithm to solve problem (QUBO-SDO) we need to choose the sets $\mathcal{P}_i$, and provide separation oracles for them. This is what we do in the following section.

7.3.2.1 Oracle Construction

The goal of Hamiltonian Updates is to solve, for fixed $\gamma \in [-1, 1]$, the following feasibility problem:

$$\begin{aligned}
\text{find} \quad &\rho \in \{X \in \mathcal{S}_+^n : \mathrm{tr}(X) = 1\} \cap \mathcal{C}_\gamma \cap \mathcal{D}_n \\
\text{where} \quad &\mathcal{C}_\gamma = \left\{ X : \mathrm{tr}\left(\widetilde{C}X\right) \geq \gamma \right\}, \\
&\mathcal{D}_n = \left\{ X : \langle i|X|i \rangle = \frac{1}{n}, i \in [n] \right\}.
\end{aligned} \tag{7.8}$$

One can observe that the set $\mathcal{C}_\gamma$ constitutes a halfspace, while $\mathcal{D}_n$ is an affine space of codimension n. The sets of observables for $\mathcal{C}_\gamma$ and $\mathcal{D}_n$ are given by $\widetilde{\mathcal{C}}_\gamma$ and $\widetilde{\mathcal{D}}_n$ respectively, with

$$\widetilde{\mathcal{C}}_\gamma = \{-\widetilde{C}\}, \text{ and } \widetilde{\mathcal{D}}_n = \{D \in \mathbb{R}^{n \times n} : \|D\| \leq 1, \ D \text{ is diagonal}\}.$$

As noted in [BKF22], it follows

$$\max_{P \in \tilde{\mathcal{C}}_\gamma} \min_{Y \in \mathcal{C}_\gamma} \operatorname{tr}(P(\rho - Y)) \leq \epsilon \iff -\operatorname{tr}\left(\tilde{C}(\rho - Y)\right) \leq \epsilon \quad \text{for some } Y \in \mathcal{C}_\gamma,$$

which in turn implies $\operatorname{tr}\left(\tilde{C}\rho\right) \geq \gamma - \epsilon$.

Given the structure of $\mathcal{C}_\gamma$ and $\mathcal{D}_n$, the authors in [BKF22] suggest the following two ϵ-separation oracles:

$O_{\mathcal{C}_\gamma}$: compute an approximation $\tilde{c}$ of $\operatorname{tr}\left(\tilde{C}\rho\right)$ up to additive error $\frac{\epsilon}{4}$. Check if $\tilde{c} \geq \gamma - \frac{3\epsilon}{4}$ and output $P = -\tilde{C}$ if the inequality is violated.

$O_{\mathcal{D}_n}$: compute an approximation $\tilde{p} \in \mathbb{R}^n$ of $p_i = \langle i|\rho|i\rangle$ satisfying $\sum_{i=1}^{n} |p_i - \tilde{p}_i| \leq \frac{\epsilon}{4}$.
Check if $\sum_{i=1}^{n} \left|\tilde{p}_i - \frac{1}{n}\right| \leq \frac{3\epsilon}{4}$ and output $P = \sum_{i=1}^{n} \left(\mathbb{I}\left\{\tilde{p}_i > \frac{1}{n}\right\} - \mathbb{I}\left\{\tilde{p}_i < \frac{1}{n}\right\}\right) |i\rangle\langle i|$ if the inequality is violated.

For any given

$$\rho_H = \frac{\exp(-H)}{\operatorname{tr}(\exp(-H))},$$

the required separation oracles are straightforward to implement on a classical computer that has access to ρ_H. Thus, classically we only need to prepare ρ_H once and store it to build the separation oracles. The next result from [BKF22] establishes that computing an $\mathcal{O}(\log(n)\epsilon^{-1})$-degree Taylor series suffices to produce accurate approximations.

Lemma 7.4 (Lemma 3.2 in [BKF22]). *Fix a Hermitian $n \times n$ matrix H, an accuracy ϵ, and let ℓ be the smallest even number satisfying $(\ell+1)(\log(\ell+1)-1) \geq 2\|H\| + \log(n) + \log\left(\frac{1}{\epsilon}\right)$. Then, the truncated matrix exponential $T_\ell = \sum_{k=0}^{\ell} \frac{1}{k!}(-H)^k$ satisfies*

$$\left\|\frac{\exp(-H)}{\operatorname{tr}\left(\exp(-H)\right)} - \frac{T_\ell}{\operatorname{tr}(T_\ell)}\right\|_{\operatorname{tr}} \leq \epsilon.$$

The task of implementing our separation oracles and testing feasibility on a quantum computer reduces to preparing Gibbs states [BKF22], which are used to test closeness to the sets $\mathcal{C}_\gamma$ and $\mathcal{D}_n$ via quantum measurements. While in Lemma 7.4 we bound the number of required Taylor series steps for computing ρ via a matrix exponential, in the quantum case we bound the number of copies of ρ required to estimate its diagonal entries and expectation values $\operatorname{tr}(A\rho)$.

Lemma 7.5. *Fix $\epsilon \in (0,1)$. Let ρ be a $\log(n)$-qubit quantum state and U a $(1, \log(n)+2, \epsilon/(2n))$-block-encoding of $\tilde{C} = C\|C\|_F^{-1}$. Then, we can implement the oracle $O_{\mathcal{C}_\gamma}$ on a quantum computer given access to $\mathcal{O}(\epsilon^{-1})$ copies of a state that is an $\frac{\epsilon}{8}$-approximation of the input state ρ in trace distance and $\mathcal{O}(\epsilon^{-1})$ applications of U and $U^\dagger$. The oracle $O_{\mathcal{D}_n}$ can be implemented using $\mathcal{O}(n\epsilon^{-2})$ $\frac{\epsilon}{8}$-approximate copies of the input, and the classical post-processing time needed to implement the oracle is $\mathcal{O}(n\epsilon^{-2})$.*

Proof. First, note that we can obtain an estimate $\tilde{p}$ of the diagonal elements of ρ whose total variation distance from p is no more than $\frac{\epsilon}{8}$ using $\tilde{\mathcal{O}}_n\left(n\epsilon^{-2}\right)$ copies of ρ to measure ρ in the computational basis. Further, provided accesses to ρ and a $(1, \log(n)+2, \epsilon/(2n))$-block-encoding U of $\widetilde{C}$, by Lemma 3.19, a trace estimator for $\mathrm{tr}\left(\widetilde{C}\rho\right)$ with bias at most $\frac{\epsilon}{n}$ can be implemented using $\tilde{\mathcal{O}}(1)$ uses of U and $U^\dagger$ and $\tilde{\mathcal{O}}_{\frac{n}{\epsilon}}(1)$ elementary operations. From here, applying amplitude estimation using $\mathcal{O}(\epsilon^{-1})$ quantum samples (i.e., state preparation unitaries) from the trace estimator to suffice to compute an approximation $\mathrm{tr}\left(\widetilde{C}\rho\right)$ up to additive $\frac{\epsilon}{8}$ to implement O_{c_γ}. The rest of the proof exactly follows the proof of [BKF22, Lemma 3.3]. $\qquad\square$

We remark that multidimensional phase estimation techniques from [vA21] could improve the dependence on ϵ^{-1} for estimating the diagonal elements of ρ to linear, which is a factor ϵ^{-1} better than a naïve application of computational basis measurements. However, in the context of the iterative refinement scheme we present later, the improvement would only reduce the amount of constant overhead in the overall running time, and multidimensional phase estimation has a larger gate complexity (which can be reduced with QRAM).

We point out that our quantum implementation of Algorithm 8 uses quantum samples to test closeness *classically* at each iterate. There are quantum approaches to performing closeness testing, see, e.g., the works of Gilyén and Li [GL19] and Li and Luo [LL23]. While this could reduce the cost of implementing the separation oracles in a quantum algorithm, it is not readily apparent how the refining methodology we propose could work without classical access to estimates of the objective value and the diagonal elements associated with the solution matrix. However, we believe iterative refinement could find wider application beyond the problems studied in this work, and we speculate that being able to perform refining steps in a fully quantum manner could be advantageous. We therefore find it important to communicate these ideas in order to stimulate investigation into this direction.

There are also numerous ways to prepare Gibbs states using a quantum computer [CS17, Fra17, KB16, PW09, vAG19, vAGGdW20, YAG12]. Following [BKF22], we utilize the Gibbs sampler from [PW09] when working with the sparse-access input model, and for the QRAM input model we consider Gibbs sampling techniques introduced in [vAG19].

7.3.3 Complexity

Having understood the cost of constructing the oracles in both the classical and quantum settings, we are now in a position to analyze the complexity associated with using Algorithm 8 to obtain solutions to (OptRelaxed) and approximations to (QUBO-SDO). Relevant to this discussion is the following result, which imposes precision requirements on solving (QUBO-SDO) to an additive error of the order $\mathcal{O}\left(n\|C\|_F\epsilon\right)$ using Algorithm 8.

Proposition 7.6 (Proposition 3.1 in [BKF22]). *Let ρ be an ϵ^4-accurate solution to the relaxed SDO problem (OptRelaxed) with input matrix C. Let $\gamma_{\epsilon^4} = \mathrm{tr}(C\rho)$ be the value attained by ρ. Then, there is a quantum state ρ^* at trace distance $\mathcal{O}(\epsilon)$ of ρ such that $n\rho^*$ is a feasible point of*

SDO problem (QUBO-SDO). *In particular*

$$\left| \gamma_{\epsilon^4} n \|C\|_F - \mathrm{tr}\left(n\rho^*C\right) \right| = \mathcal{O}\left(n\|C\|_F \epsilon\right).$$

Moreover, it is possible to construct ρ^* in time $\mathcal{O}(n^2)$ given the entries of ρ.

We do not provide a proof of this result here, as later we will provide an improved approximation guarantee and a proof of the improved statement.

7.3.3.1 Classical running time

Using Lemma 7.4 in combination with Theorem 7.3, we can bound the running time required to solve (OptRelaxed) to additive error ϵ using a classical implementation of Algorithm 8.

Proposition 7.7. *Suppose that C has row sparsity s. Then, the classical cost of solving* (OptRelaxed) *up to additive error ϵ using Algorithm 8 is $\mathcal{O}\left(\min\{n^2 s, n^\omega\}\log^2(n)\epsilon^{-3}\right)$.*

Proof. The result follows directly from the proof of Corollary 3.1 in [BKF22], but we repeat the argument here for completeness.

First, observe that over the course of the iterations $t = 0, \ldots, T$, the operator norms $\|H^{(t)}\|$ do not become prohibitively large. This follows from initializing $H^{(0)} = \mathbf{0}^{n\times n}$, and that by (7.6), the inequality

$$\left\|H^{(t+1)} - H^{(t)}\right\| \le \frac{\epsilon}{16} \left\|P^{(t)}\right\| \le \frac{\epsilon}{16}$$

holds for all t. By Theorem 7.3, Algorithm 8 requires at most $T = \lceil 64\log(n)\epsilon^{-2}\rceil$ iterations, which implies $\|H^{(t)}\| \le 4\log(n)\epsilon^{-1}$ for all t.

By Lemma 7.4, it suffices to compute $\mathcal{O}(\log(n)\epsilon^{-1})$ steps of the Taylor series corresponding to $\exp(-H^{(t)})$ in order to obtain a matrix $\tilde{\rho}^{(t)}$ that is at most a trace distance of $\frac{\epsilon}{4}$ from $\rho^{(t)}$. Moreover, given that $H^{(t)}$ is defined as a linear combination of $\widetilde{C}$ with a diagonal matrix, matrix multiplication involving $H^{(t)}$ can be carried out in $\mathcal{O}(\min\{n^2 s, n^\omega\})$ arithmetic operations. Given classical access to $\tilde{\rho}^{(t)}$, the diagonal constraints comprising $\mathcal{D}_n$ can be checked in time $\mathcal{O}(n)$, whereas computing $\mathrm{tr}\left(\widetilde{C}\tilde{\rho}^{(t)}\right)$ requires $\mathcal{O}(ns)$ arithmetic operations. Thus, the dominant operation at each iteration is computing the matrix exponential and the classical per-iteration cost of Algorithm 8 is given by

$$\mathcal{O}\left(\min\{n^2 s, n^\omega\}\log(n)\epsilon^{-1}\right).$$

Taking into account the iteration bound $\mathcal{O}(\log(n)\epsilon^{-2})$ provided in Theorem 7.3, we arrive at an overall running time of

$$\mathcal{O}\left(\min\{n^2 s, n^\omega\}\log^2(n)\epsilon^{-3}\right).$$

The proof is complete. $\qquad\square$

The next corollary from [BKF22] follows from Proposition 7.6 in the context of the previous result, and provides the overall running time of Algorithm 8 to solve (QUBO-SDO) to additive error $\mathcal{O}\left(n\|C\|_F\epsilon\right)$ in the classical setting.

Corollary 7.1. *Suppose that C has row-sparsity s. Then, the classical cost of solving (QUBO-SDO) up to an additive error $\mathcal{O}\left(n\|C\|_F\epsilon\right)$ using Algorithm 8 is $\mathcal{O}\left(\min\{n^2s, n^\omega\}\log^2(n)\epsilon^{-12}\right)$.*

Proof. By Proposition 7.7, Algorithm 8 requires time

$$\mathcal{O}\left(\min\{n^2s, n^\omega\}\log^2(n)\tilde{\epsilon}^{-3}\right),$$

to solve (OptRelaxed) up to additive error $\tilde{\epsilon}$. In order to satisfy the approximation guarantee for (QUBO-SDO) given in Proposition 7.6, it suffices to solve (OptRelaxed) to error $\tilde{\epsilon} = \epsilon^4$. Plugging in this value for the precision parameter, the total cost required to solve (QUBO-SDO) up to an additive error $\mathcal{O}\left(n\|C\|_F\epsilon\right)$ using Algorithm 8 is

$$\mathcal{O}\left(\min\{n^2s, n^\omega\}\log^2(n)\tilde{\epsilon}^{-3}\right) = \mathcal{O}\left(\min\{n^2s, n^\omega\}\log^2(n)(\epsilon^4)^{-3}\right)$$
$$= \mathcal{O}\left(\min\{n^2s, n^\omega\}\log^2(n)\epsilon^{-12}\right).$$

$\square$

7.3.3.2 Quantum running time

Combining the sampling requirements provided in Lemma 7.5 with the cost of preparing a single Gibbs state and the iteration bound from Theorem 7.3 gives the complexity of Algorithm 8 when run on a quantum computer. However, Gibbs samplers based on the block-encoding framework depend only poly-logarithmically on the inverse precision, therefore they are exponentially faster (in the parameter ϵ^{-1}) compared to the Gibbs sampling algorithm from [PW09] utilized in [BKF22]. It thus makes sense to analyze the running time in the more efficient model. This will require an efficient data structure for storing y so that we can efficiently prepare linear combinations of block-encodings.

Lemma 7.8 (Lemma 15 in [vAG19]). *There is a data structure that can store an m-dimensional χ-sparse vector y with θ-precision using a QRAM of size $\widetilde{\mathcal{O}}_{\frac{m}{\theta}}(\chi)$. Furthermore:*

- *Given a classical $\mathcal{O}(1)$-sparse vector, adding it to the stored vector has classical cost $\widetilde{\mathcal{O}}_{\frac{m}{\theta}}(1)$.*

- *Given that $\beta \geq \|y\|_1$, we can implement a (symmetric) $(\beta, \widetilde{\mathcal{O}}_{\frac{m}{\theta}}(1), \theta)$-state preparation pair for y with $\widetilde{\mathcal{O}}_{\frac{m}{\theta}}(1)$ queries to the QRAM.*

Corollary 7.2 (Corollary 16 in [vAG19]). *Suppose $A_1, \ldots, A_m$ are Hermitian matrices with operator norm at most 1, and that $y \in \mathbb{R}^m$ satisfies $\|y\|_1 \leq \beta$. Having access to the above data structure for y, we can prepare one copy of the Gibbs state*

$$\rho = \frac{\exp\left(-\sum_{i=1}^m y_i A_i\right)}{\mathrm{tr}\left(\exp\left(-\sum_{i=1}^m y_i A_i\right)\right)}$$

using $\widetilde{\mathcal{O}}_\theta(\sqrt{n}\alpha\beta)$ accesses to the data structure for y and block-encodings of $A_1,\dots,A_m$.

We can now use Corollary 7.2 in combination with results from Sections 2.5.6 and 3.4.1 to establish the running time of Algorithm 8 in the QRAM input model.

Proposition 7.9. *Let $\widetilde{C} = C\|C\|_F^{-1} \in \mathcal{S}^n$ be stored in QRAM. Then, the complexity of solving* (OptRelaxed) *up to additive error ϵ with Algorithm 8 using the QRAM input model is*

$$\widetilde{\mathcal{O}}_{\frac{n}{\epsilon}}\left(n^{1.5}\epsilon^{-5}\right).$$

Here, the complexity corresponds to the number of accesses to the QRAM.

Proof. Given that $\widetilde{C}$ is stored in QRAM, Lemma 3.3*(ii)* asserts that when constructing a block-encoding of $\widetilde{C}$, one can set the subnormalization factor to be $\alpha_C = \left\|\widetilde{C}\right\|_F = 1$. Hence, one can construct a $(1, \log(n) + 2, \epsilon/(2n))$-block-encoding of $\widetilde{C}$ in time $\widetilde{\mathcal{O}}_{\frac{n}{\epsilon}}(1)$.

Next, recall that in iteration $t \in [T]$ of Algorithm 8, our Hamiltonian is defined as

$$H^{(t)} = y_1^{(t)}\widetilde{C} + y_2^{(t)}D^{(t)},$$

where $D^{(t)}$ is a diagonal matrix with the diagonal entries taking value $-1, 0$ or 1. The diagonal elements of D change in each iteration, and therefore, a new D must be block-encoded in each iteration. For this, we use the QRAM model described in Section 2.5.5, which allows for insertions to be made in time $\widetilde{\mathcal{O}}_n(1)$ to keep the cost of this step negligible. Provided a classical description of D, we can store D in the QRAM in time $\mathcal{O}(n\log(n))$. Applying Lemma 3.2, a $(1, \log(n) + 3, \epsilon)$-block-encoding of $D^{(t)}$ can be constructed in time $\widetilde{\mathcal{O}}_{\frac{n}{\epsilon}}(1)$.

In an earlier discussion we saw that any y obtained from a call to Algorithm 8 will satisfy $\|y\|_1 = \widetilde{\mathcal{O}}_n(\epsilon^{-1})$ if we call Algorithm 8 using precision ϵ (see, e.g., equation (7.7)). Hence, an application of Corollary 7.2 with $\beta = \widetilde{\mathcal{O}}_n(\epsilon^{-1})$ implies that we can prepare one copy of our Gibbs state using

$$\widetilde{\mathcal{O}}_{\frac{n}{\epsilon}}\left(\sqrt{n}\alpha\epsilon^{-1}\right)$$

accesses to the data structure for y and the block-encodings of $\widetilde{C}$ and D, where α is defined as the maximum over the subnormalization factors used to block-encode $\widetilde{C}$ and D. Since $\alpha = \max\{\alpha_C, \alpha_D\} = 1$, it follows

$$\widetilde{\mathcal{O}}_{\frac{n}{\epsilon}}\left(\sqrt{n}\alpha\epsilon^{-1}\right) = \widetilde{\mathcal{O}}_{\frac{n}{\epsilon}}\left(\sqrt{n}\epsilon^{-1}\right).$$

Now, one can see from Lemma 7.5 that the cost of constructing O_{D_n} dominates that of constructing O_{C_γ}. Noting that O_{D_n} can be implemented using $\mathcal{O}(n\epsilon^{-2})$ copies of a state that is an $\frac{\epsilon}{8}$-approximation of the input state ρ in trace distance and its inverse, the per-iteration cost of Algorithm 8 in the QRAM input model is given by

$$\widetilde{\mathcal{O}}_{\frac{n}{\epsilon}}\left(n^{1.5}\epsilon^{-3}\right).$$

Factoring in the iteration bound of $\tilde{\mathcal{O}}_n(\epsilon^{-2})$ from Theorem 7.3, it follows that when provided access to QRAM, Algorithm 8 solves (OptRelaxed) up to additive error ϵ using

$$T_{HU}^{\mathrm{quantum}} = \tilde{\mathcal{O}}_{\frac{n}{\epsilon}}\left(n^{1.5}\epsilon^{-5}\right)$$

accesses to the QRAM. The proof is complete. $\qquad\square$

Corollary 7.3. *Let $\tilde{C} \in \mathcal{S}^n$ be stored in QRAM. Then, the complexity of solving* (QUBO-SDO) *up to additive error $\mathcal{O}(n\|C\|_F\epsilon)$ with Algorithm 8 using the QRAM input model is*

$$\tilde{\mathcal{O}}_{\frac{n}{\epsilon}}\left(n^{1.5}\epsilon^{-20}\right).$$

Here, the complexity corresponds to the number of accesses to the QRAM.

Proof. By Proposition 7.9, Algorithm 8 requires

$$\tilde{\mathcal{O}}_{\frac{n}{\tilde{\epsilon}}}\left(n^{1.5}\tilde{\epsilon}^{-5}\right),$$

accesses to the QRAM to solve (OptRelaxed) up to additive error $\tilde{\epsilon}$. In order to satisfy the approximation guarantee for (QUBO-SDO) given in Proposition 7.6, it suffices to solve (OptRelaxed) to error $\tilde{\epsilon} = \epsilon^4$. Plugging in this value for the precision parameter, the total cost required to solve (QUBO-SDO) up to an additive error $\mathcal{O}\left(n\|C\|_F\epsilon\right)$ using Algorithm 8 is

$$\tilde{\mathcal{O}}_{\frac{n}{\tilde{\epsilon}}}\left(n^{1.5}\tilde{\epsilon}^{-5}\right) = \tilde{\mathcal{O}}_{\frac{n}{\epsilon}}\left(n^{1.5}(\epsilon^4)^{-5}\right) = \tilde{\mathcal{O}}_{\frac{n}{\epsilon}}\left(n^{1.5}\epsilon^{-20}\right).$$

The proof is complete. $\qquad\square$

Corollary 7.3 establishes that utilizing Gibbs samplers and trace estimators based on the block-encoding framework for our oracle construction in Algorithm 8 leads to an

$$\mathcal{O}\left(\sqrt{s}^{-1+o(1)}\epsilon^{-8+o(1)}\exp\left(1.6\sqrt{\log(\epsilon^{-4})}\right)\right)$$

speedup over the running time result provided in [BKF22, Corollary 3.2] when applied to solving (QUBO-SDO). Yet, the costly accuracy requirements for the rounding procedure (see, e.g., Proposition 7.6) lead to a prohibitive scaling in the inverse precision for the overall running time. Given the advantageous dependence on the dimension, as compared to classical algorithms, we study how to improve the dependence on the precision parameter. This is discussed next.

7.4 Iterative Refinement for SDO approximations of QU-BOs

In this section, we introduce an iterative refinement method for obtaining accurate solutions to the renormalized relaxed SDO problem (OptRelaxed), that at a high level can be viewed as solving a series of problems related to the *feasibility problem* (7.8) associated with (OptRelaxed). We then discuss how to test ϵ-closeness to the convex sets which comprise the feasible regions of the intermediate refining problems before presenting our algorithm in full detail. We conclude the section by proving our algorithm's correctness and iteration complexity, and use these results to provide an improved approximation guarantee.

7.4.1 The refining problem

To develop an iterative refinement scheme for (OptRelaxed), we need to design a problem whose solution can be used to improve the quality of solutions to (OptRelaxed). Suppose we run Algorithm 8 and obtain an ϵ-precise solution $\tilde{\rho}$ to (OptRelaxed). Letting $\tilde{\gamma} = \operatorname{tr}\left(\widetilde{C}\tilde{\rho}\right)$, $\tilde{\rho}$ must satisfy

$$\operatorname{tr}\left(\widetilde{C}\tilde{\rho}\right) = \tilde{\gamma} \geq \gamma - \epsilon,$$

$$\sum_{i=1}^{n}\left|\langle i|\tilde{\rho}|i\rangle - \frac{1}{n}\right| \leq \epsilon.$$

In *refining* our solution to (OptRelaxed), we should aim to reduce the total variation distance from the distribution along the diagonal elements of our solution to the uniform distribution, while also improving the precision to which the optimal objective value is approximated. Thus, an improved solution ρ' should obey

$$\operatorname{tr}\left(\widetilde{C}\rho'\right) \geq \gamma - \epsilon',$$

$$\sum_{i=1}^{n}\left|\langle i|\rho'|i\rangle - \frac{1}{n}\right| \leq \epsilon',$$

with $\epsilon' < \epsilon$. The basic idea behind constructing the refining problem is to use our current solution $\tilde{\rho}$ to first shift the renormalized relaxed SDO problem (OptRelaxed) to the origin, and then scale the shifted problem back to the domain of the original problem. In particular, we solve a series of problems related to the feasibility problem (7.8).

Let $\varepsilon \in \mathbb{R}^n$ be a vector whose elements are the residuals along the diagonal $\varepsilon_i = \tilde{\rho}_{ii} - \frac{1}{n}$ for $i \in [n]$, and $\eta \geq 1$ to be a scalar defined as

$$\eta = \frac{1}{\max\left\{\gamma - \operatorname{tr}\left(\widetilde{C}\tilde{\rho}\right), \sum_{i=1}^{n}|\varepsilon_i|\right\}} = \frac{1}{\max\left\{\gamma - \operatorname{tr}\left(\widetilde{C}\tilde{\rho}\right), \left\|\sum_{i\in[n]} \langle i|\tilde{\rho}|i\rangle|i\rangle\langle i| - n^{-1}I\right\|_{\operatorname{tr}}\right\}}.$$

Using these quantities, the *refining problem* is given by:

$$\text{find} \quad \rho^r \in \{X \in \mathcal{S}_+^n : \text{tr}(X) = 1\} \cap \mathcal{C}_{\eta(\gamma - \tilde{\gamma})} \cap \mathcal{D}_{\eta\varepsilon}$$

$$\text{where} \quad \mathcal{C}_{\eta(\gamma - \tilde{\gamma})} = \left\{ X : \text{tr}\left(\tilde{C}(Q \circ X) \right) \geq \eta(\gamma - \tilde{\gamma}) \right\}, \tag{7.9}$$

$$\mathcal{D}_{\eta\varepsilon} = \left\{ X : \langle i|X|i \rangle = \eta|\varepsilon_i|, \ \forall i \in [n] \right\},$$

where $Q \in \mathcal{S}^n$ is a matrix whose diagonal elements are chosen such that for any $X \in \mathcal{D}_{\eta\varepsilon}$, we have

$$(Q \circ X)_{ii} = \text{sign}(-\varepsilon_i)\eta|\varepsilon_i|$$

for $i \in [n]$. Further details and requirements on the structure of Q are specified later in this section. We refer to solutions ρ^r to (7.9) as *refining solutions*, which we use to update our current solution $\tilde{\rho}$ to (OptRelaxed).

The set $\mathcal{D}_{\eta\varepsilon}$ is comprised of the diagonal constraints

$$\langle i|X|i \rangle = \eta|\varepsilon_i|, \quad \forall i \in [n],$$

and similar to $\mathcal{D}_n$, is an affine space with codimension n. Our use of the absolute value function of the residuals and scaling by η ensures the viability of applying Gibbs sampling techniques to solve the refining problem (7.9); the diagonal terms of any density matrix must be nonnegative and sum to 1. Whenever

$$\sum_{i=1}^{n} |\varepsilon_i| > \gamma - \text{tr}\left(\tilde{C}\tilde{\rho} \right),$$

then $\eta\|\varepsilon\|_1 = 1$, and the parameter η therefore scales the shifted problem back to the space of the $\log(n)$-qubit mixed states, ensuring that any solution ρ^r to (7.9) is indeed a (trace normalized) Gibbs state.

On the other hand, should it be the case that

$$\sum_{i=1}^{n} |\varepsilon_i| \leq \gamma - \text{tr}\left(\tilde{C}\tilde{\rho} \right),$$

then for any $X \in \mathcal{D}_{\eta\varepsilon}$ we have $\text{tr}(X) \leq 1$, rather than $\text{tr}(X) = 1$. Our primal SDO oracle in Algorithm 8 solves feasibility problems in which the trace upper bound is tight, i.e., $\text{tr}(X) = 1$. The authors in [vAG19] note that this can be dealt with adding one extra variable w such that

$$\bar{\rho}^r := \begin{bmatrix} \rho^r & 0 \\ 0 & w \end{bmatrix}.$$

Then, $\text{tr}(\bar{\rho}^r) = 1$ and $\bar{\rho}^r \succeq 0$ imply that $\text{tr}(\rho^r) \leq 1$, and as a result we obtain an SDO problem that is equivalent to (7.9). Since we know exactly the amount of subnormalization, we can also get rid of the extra variable in subsequent calculations and re-scale the trace back to 1 when necessary (e.g., when combining solutions from multiple iterative refinement iterations for trace estimations). Crucially, using the input models described in Section 2.5.6, these modifications do not introduce more than constant overhead in the overall complexity, as the problem data in this case is simply given by

$$\overline{C} = \begin{bmatrix} \tilde{C} & 0 \\ 0 & 0 \end{bmatrix}, \quad \overline{Q} = \begin{bmatrix} Q & 0 \\ 0 & 0 \end{bmatrix},$$

with $(\overline{C}, \overline{Q}) \in \mathcal{S}^{n+1} \times \mathcal{S}^{n+1}$.

The Hadamard product $Q \circ \rho^r$ that appears in the definition of $\mathcal{C}_{\eta(\gamma - \tilde{\gamma})}$ is required for similar reasons; properly setting Q allows us to drive the total variation distance from the distribution along the diagonal elements of our solution to the uniform distribution to zero using the solutions to the refining problem. Later, in Section 7.4.3 we demonstrate that this can be achieved by generating a sequence of iterates $\tilde{\rho}, \hat{\rho}$, where we obtain $\hat{\rho}$ from $\tilde{\rho}$ using the rule

$$\hat{\rho} = \tilde{\rho} + \frac{1}{\eta} Q \circ \rho^r, \tag{7.10}$$

with a suitable choice for Q being

$$Q = (ee^\top - I) + \mathrm{diag}\left(\mathrm{sign}(-\varepsilon)\right) = \begin{pmatrix} \mathrm{sign}(-\varepsilon_1) & 1 & \cdots & 1 \\ 1 & \mathrm{sign}(-\varepsilon_2) & \ddots & \vdots \\ \vdots & \ddots & \ddots & 1 \\ 1 & \cdots & 1 & \mathrm{sign}(-\varepsilon_n) \end{pmatrix}. \tag{7.11}$$

Choosing Q in this manner also implies that the Hadamard product $Q \circ A$ can be carried out classically using $\mathcal{O}(n)$ arithmetic operations for any $A \in \mathbb{R}^{n \times n}$, as the element-wise products $Q_{ij}A_{ij} = A_{ij}$ for $i \neq j$. Similarly, updating Q at each iterate only requires updating its diagonal elements, an $\mathcal{O}(n)$ operation.

It is important to note that the update we propose in (7.10) does not preserve positive semidefiniteness in general. However, later in our analysis, we demonstrate that the eigenvalues of the updated solution $\hat{\rho}$ are only slightly negative in the worst case, i.e., $\lambda_{\min}(\hat{\rho}) \geq -\delta$ for a value $\delta > 0$ that gets progressively smaller over the course of the algorithm; one can restore positive semidefiniteness by adding δ to the diagonal elements of the final solution, and we renormalize the trace by $(1 + n\delta)$. We show that these modifications required to restore positive semidefiniteness have only a mild (in fact, constant) impact on feasibility. To this end, we will bound the eigenvalues of Q. We first state a special instance of Weyl's inequality.

Lemma 7.10. *Suppose that $A \in \mathbb{R}^{n \times n}$ and $B \in \mathbb{R}^{n \times n}$ are Hermitian matrices. Then*

$$\lambda_{\min}(A + B) \geq \lambda_{\min}(A) + \lambda_{\min}(B).$$

Using the preceding lemma, the following result bounds the minimum eigenvalue of Q.

Lemma 7.11. *Suppose that $Q \in \mathcal{S}^n$ is defined according to Equation (7.11). Then, $\lambda_{\min}(Q) \geq$*

-2.

Proof. Let $A = (ee^\top - I)$ and $B = \mathrm{diag}\,(\mathrm{sign}(-\varepsilon))$, such that $Q = A + B$. Now, it can be easily seen from the definition of A that $A + I$ is an all-ones matrix of dimension n. Upon performing row-reduction (via, e.g., Guassian elimination) on A, it is trivial to observe that the resulting row-echelon form will have $n-1$ zero rows, and as a consequence, A has the eigenvalue -1, repeated (at least) $n-1$ times. Further, since $\mathrm{tr}\,(A) = 0$, the other eigenvalue is $n-1$. Therefore, we have $\lambda_{\min}(A) \geq -1$. On the other hand, B is a diagonal matrix whose diagonal elements can take value -1, 0, or 1, from which $\lambda_{\min}(B) \geq -1$ readily follows.

Applying Lemma 7.10, we obtain

$$\lambda_{\min}(Q) = \lambda_{\min}(A + B) \geq \lambda_{\min}(A) + \lambda_{\min}(B) \geq -2.$$

The proof is complete. $\qquad\square$

7.4.2 Oracle construction for the refining problem

In order to construct separation oracles for testing closeness to $\mathcal{C}_{\eta(\gamma - \tilde{\gamma})}$, we rely on the following result.

Lemma 7.12. *Let E, F and $G \in \mathcal{S}^n$. We have*

$$\mathrm{tr}\,\big(G(E \circ F)\big) = \mathrm{tr}\,\big((E \circ G)F\big).$$

Proof. Applying Lemma 7.1 with $m = n$, we have

$$\big[(E \circ F)G\big]_{ii} = \big[(E \circ G)F\big]_{ii} \quad \forall i \in [n].$$

Note that we have dropped the transpose terms, as E, F and G are symmetric matrices, and hence, so are $E \circ F$ and $E \circ G$. It follows

$$\mathrm{tr}\,\big(G(E \circ F)\big) = \mathrm{tr}\,\big((E \circ F)G\big) = \sum_{i \in [n]} \big[(E \circ F)G\big]_{ii} = \sum_{i \in [n]} \big[(E \circ G)F\big]_{ii} = \mathrm{tr}\,\big((E \circ G)F\big).$$

$\qquad\square$

In addition to $Q \in \mathcal{S}^n$, we also require $\max_{i,j \in [n]}\{|Q_{ij}|\} \leq 1$ to avoid any normalization issues with respect to $Q \circ \tilde{C}$. Note that defining of Q according to equation (7.11) satisfies both of these properties trivially, as each of the diagonal elements are 1, 0, or -1, while the off-diagonal elements are all set to 1. This idea is formalized next.

Lemma 7.13. *Let $A, Q \in \mathcal{S}^n$ with $\max_{i,j \in [n]}\{|Q_{ij}|\} \leq 1$ and $\|A\|_F \leq 1$, then,*

$$\|Q \circ A\| \leq 1.$$

Proof. Under the stated conditions for Q, it follows

$$\|Q \circ A\|_F^2 = \sum_{i \in [n]} \sum_{j \in [n]} \left([Q \circ A]_{ij}\right)^2 = \sum_{i \in [n]} \sum_{j \in [n]} \left(Q_{ij} \cdot A_{ij}\right)^2 = \sum_{i \in [n]} \sum_{j \in [n]} \left(Q_{ij}\right)^2 \left(A_{ij}\right)^2$$

$$\leq \sum_{i \in [n]} \sum_{j \in [n]} \left(A_{ij}\right)^2 = \|A\|_F^2,$$

and applying the square root throughout the above we obtain $\|Q \circ A\|_F \leq \|A\|_F$. From here, the result follows upon noting $\|A\|_F = 1$ and $\|A\| \leq \|A\|_F$ is true for all $A \in \mathbb{R}^{n \times n}$. $\qquad\square$

Although the sets $\mathcal{C}_\gamma$ and $\mathcal{D}_n$ differ from their refining counterparts $\mathcal{C}_{\eta(\gamma-\tilde{\gamma})}$ and $\mathcal{D}_{\eta\varepsilon}$, their dissimilarity merely affects the right hand side of the inequality defining the sets, and are thus no more difficult to construct. Just as in the case of (7.8), the task of obtaining separation oracles for the refining problem (7.9) in the quantum regime reduces to preparing many copies of Gibbs states. Likewise, these oracles can also be implemented on a classical computer, given access to ρ^r.

The similarities between (7.8) and (7.9) become transparent when we demonstrate that they are specific instances of the same problem. In particular, it is easy to see that solving (7.8) corresponds to solving

$$\text{find} \quad \rho \in \{X \in \mathcal{S}_+^n : \operatorname{tr}(X) = 1\} \cap \mathcal{C}_{\eta(\gamma-\tilde{\gamma})} \cap \mathcal{D}_{\eta\varepsilon}$$

$$\text{where} \quad \mathcal{C}_{\eta(\gamma-\tilde{\gamma})} = \left\{ X : \operatorname{tr}\left(\tilde{C}Q \circ X\right) \geq \eta(\gamma - \tilde{\gamma}) \right\}, \qquad \text{(RefProb)}$$

$$\mathcal{D}_{\eta\varepsilon} = \left\{ X : \langle i|X|i\rangle = \eta|\varepsilon_i|, \ \forall i \in [n] \right\},$$

with $\varepsilon_i = \frac{1}{n}$, $\eta = 1$, $Q = ee^\top$, and $\tilde{\gamma} = 0$. In view of this relationship, we can unify the oracle construction for (7.8) and (7.9) as follows:

$O_{\mathcal{C}_{\eta(\gamma-\tilde{\gamma})}}$: Compute an approximation $\tilde{c}$ of $\operatorname{tr}\left(Q \circ \tilde{C}\rho\right)$ up to additive error $\frac{\epsilon}{4}$.
Check if $\tilde{c} \geq \eta(\gamma - \tilde{\gamma}) + \dfrac{3\epsilon}{4}$ and output $P = -Q \circ \tilde{C}$ if the inequality is violated.

$O_{\mathcal{D}_{\eta\varepsilon}}$: Compute an approximation $\tilde{p} \in \mathbb{R}^n$ of $p_i = \langle i|\rho|i\rangle$ satisfying $\sum_{i \in [n]} |p_i - \tilde{p}_i| \leq \dfrac{\epsilon}{4}$.
Check if $\sum_{i \in [n]} \left|\tilde{p}_i - \eta|\varepsilon_i|\right| \leq \dfrac{3\epsilon}{4}$ and output $P = \sum_{i \in [n]} \left(\mathbb{I}\{\tilde{p}_i > \eta|\varepsilon_i|\} - \mathbb{I}\{\tilde{p}_i < \eta|\varepsilon_i|\}\right) |i\rangle\langle i|$
if the inequality is violated.

Again, the sets of observables for $\mathcal{C}_{\eta(\gamma-\tilde{\gamma})}$ and $\mathcal{D}_{\eta\varepsilon}$ are given by

$$\tilde{\mathcal{C}}_{\eta(\gamma-\tilde{\gamma})} = \{-Q \circ \tilde{C}\}, \text{ and } \tilde{\mathcal{D}}_{\eta\varepsilon} = \{D \in \mathbb{R}^{n \times n} : \|D\| \leq 1, D \text{ is diagonal}\}.$$

Although these observations are straightforward, they justify our use of Algorithm 8 as a semidefinite optimization oracle that solves a convex feasibility problem at hand in every iteration for different values of Q. In particular, these facts, along with Lemmas 7.12 and 7.13 ensure that the complexity results in Propositions 7.7 and 7.9 hold when applying Algorithm 8 to solve (RefProb).

Proposition 7.14. *Let $Q \circ \widetilde{C} \in \mathcal{S}^n$ be stored in QRAM. Algorithm 8 solves* (RefProb) *up to additive error ϵ using*

$$\widetilde{\mathcal{O}}_{\frac{n}{\epsilon}}\left(n^{1.5}\epsilon^{-5}\right)$$

accesses to the QRAM.

Proof. Given that $Q \circ \widetilde{C}$ is stored in QRAM, Lemma 3.3*(ii)* asserts that when constructing a block-encoding of $Q \circ \widetilde{C}$, one can set the subnormalization factor to be $\alpha_C = \left\|Q \circ \widetilde{C}\right\|_F$. In particular, one can always choose $\alpha_C = 1$, as it can be seen from the proof of Lemma 7.13 that the inequality

$$\left\|Q \circ \widetilde{C}\right\|_F \leq \left\|\widetilde{C}\right\|_F = 1$$

always holds for any Q defined according to equation (7.11). Collecting these facts, one can construct a $(1, \mathcal{O}(\log(n)), \epsilon/(2n))$-block-encoding of $Q \circ \widetilde{C}$ in time $\widetilde{\mathcal{O}}_{\frac{n}{\epsilon}}(1)$. Note that the quantity $Q \circ \widetilde{C}$ remains unchanged for the duration of Algorithm 8. From here, the rest of the proof follows exactly that of Proposition 7.9 upon replacing $\widetilde{C}$, $O_{\mathcal{C}_\gamma}$ and $O_{\mathcal{D}_n}$ with $Q \circ \widetilde{C}$, $O_{\mathcal{C}_{\eta(\gamma-\hat{\gamma})}}$ and $O_{\mathcal{D}_{\eta\varepsilon}}$, respectively, in what remains. $\qquad\square$

Before proceeding further, we establish that the eigenvalues of the updated solution will never fall significantly below zero by deriving a lower bound on the minimum eigenvalue of the terms $\frac{1}{\eta}Q \circ \rho$ that are used to update the overall solution in each iteration of our refinement scheme according to (7.10).

Proposition 7.15. *Let ρ be a solution to* (RefProb) *obtained from running Algorithm 8 using precision $\epsilon \in (0,1)$. Then,*

$$\frac{1}{\eta}Q \circ \rho \succeq -2 \cdot \left(\|\varepsilon\|_1 + \frac{\epsilon}{\eta}\right)n^{-1}I.$$

Proof. In what follows, we assume without loss of generality that Q has at least one negative eigenvalue (otherwise, $Q \circ \rho \succeq 0$ trivially holds), so applying Lemma 7.11 we can let $\lambda_{\min}(Q) \geq -2$. Applying Lemma 7.2, we can lower bound the minimum eigenvalue of the Hadamard product $Q \circ \rho$ as follows

$$\lambda_{\min}(Q \circ \rho) \geq \min_{i \in [n]} \rho_{ii} \cdot \lambda_{\min}(Q) \geq -2 \min_{i \in [n]} \rho_{ii}.$$

Therefore, in order to derive a worst case lower bound on $\lambda_{\min}(Q \circ \rho)$, it suffices to determine

$$\max_{\rho \in \mathcal{D}_{\eta\varepsilon}} \min_{i \in [n]} \rho_{ii}.$$

The definition of $\mathcal{D}_{\eta\varepsilon}$ asserts that when $O_{\mathcal{D}_{\eta\varepsilon}}$ is queried with precision ϵ, the diagonal elements of ρ are nonnegative and must satisfy the following:

$$\sum_{i \in [n]} \left|\rho_{ii} - \eta|\varepsilon_i|\right| \leq \epsilon, \quad \sum_{i \in [n]} \rho_{ii} \leq \eta\|\varepsilon\|_1 + \epsilon.$$

Hence, $\max_{\rho \in \mathcal{D}_{\eta\epsilon}} \min_{i \in [n]} \rho_{ii} \leq \frac{\eta \|\epsilon\|_1 + \epsilon}{n}$, and the proof is complete. $\qquad\square$

7.4.3   Iterative Refinement using Hamiltonian Updates

We are now in a position to provide our iterative refinement method for SDO approximations of QUBOs presented in full detail in Algorithm 9.

The algorithm takes three parameters as input; *(i)* ξ, the fixed (constant) precision used to test closeness to the sets $\mathcal{C}_{\eta(\gamma-\tilde{\gamma})}$ and $\mathcal{D}_{\eta\epsilon}$ in every iteration, *(ii)* ζ, the precision to which the final solves (OptRelaxed), and *(iii)* ϵ, the additive error to which we seek to solve (QUBO-SDO). In our initialization steps we set the values of Q, ϵ and η such that the call to Algorithm 8 corresponds to solving the original feasibility problem (7.8).

In each iteration k, Algorithm 9 calls Algorithm 8 with separation oracles $O_{\mathcal{C}_{\eta(\gamma-\tilde{\gamma})}}$ and $O_{\mathcal{D}_{\eta\epsilon}}$ using fixed precision ξ^2 such that every call to Algorithm 8 produces a ξ^2-precise solution $\rho^{(k)}$ to (RefProb). Using $\rho^{(k)}$ to update our solution according to (7.10) may cause us to obtain matrices $\hat{\rho}^{(k)}$ with eigenvalues that are, in the worst case, slightly negative. A shift of the spectrum defined via the bound on the minimum eigenvalue of the update term $\frac{1}{\eta^{(k)}} Q \circ \rho^{(k)}$ provided in Proposition 7.15 suffices to obtain a positive semidefinite matrix $\tilde{\rho}^{(k)}$, and we will demonstrate that it does not change the constraint violation or the objective function value by a large amount.

If $\tilde{\rho}$ is indistinguishable up to precision ζ from the maximally mixed state $n^{-1}I$ upon measurement in the computational basis, and satisfies $\mathrm{tr}\left(\widetilde{C}\tilde{\rho}\right) \geq \gamma - \zeta$, the algorithm terminates and reports $\tilde{\rho}$. Otherwise, we shift the spectrum as described above, we construct the refining problem associated with our current solution, and proceed to the next iteration. To define the parameters for the next refining problem, we first calculate the deviation of the diagonal elements from $\frac{1}{n}$, and the violation with respect to satisfying our objective value. Then, we define our scaling factor to be the reciprocal of the maximum over the ℓ_1-norm of the diagonal deviations, and the objective violation. We stress that ξ is a (chosen) constant, and does not change throughout the algorithm.

We now state a series of results in order to bound the iteration complexity of Algorithm 9, and use our findings to improve the approximation guarantee given in Proposition 7.6. We begin by proving establishing that the iterates generated by Algorithm 9 are increasingly accurate solutions to (OptRelaxed).

Theorem 7.16. *Fix a constant $\xi \in \left(0, \frac{1}{2}\right)$ and define $\tilde{\xi} = \frac{\xi}{4}$. Let $\rho^{(k)}$ be a solution to (RefProb) obtained from running Algorithm 8 using fixed precision $\tilde{\xi}^2$ in iteration k of Algorithm 9. Then, the following hold:*

(a) For $k \geq 0$, $\eta^{(k)} \geq \frac{1}{2\xi^k}$.

Algorithm 9 Iterative Refinement for SDO Approximations of QUBOs
---

Input: Error tolerances $\epsilon \in (0,1)$ and $\zeta = \left(\frac{\epsilon}{n\|C\|_F}\right)^4$, upper bound on objective value $\gamma \in [-1,1]$

Output: A matrix $\tilde{\rho} \in \{X \in \mathcal{S}_+^n : \mathrm{tr}(X) \leq 1 + \zeta\}$ satisfying

$$\max\left\{\gamma - \mathrm{tr}\left(\widetilde{C}\tilde{\rho}\right), \left\|\sum_{i\in[n]} \langle i|\tilde{\rho}|i\rangle |i\rangle\langle i| - n^{-1}I\right\|_{\mathrm{tr}}\right\} \leq \zeta$$

Initialize: $\tilde{\rho}, \hat{\rho} \leftarrow 0^{n\times n}$, $Q \leftarrow ee^\top$, $\varepsilon_i = \frac{1}{n}$ for $i \in [n]$, $\tilde{\gamma}, \hat{\gamma} \leftarrow 0$, $\eta^{(0)} \leftarrow 1$, $k \leftarrow 1$

$\tilde{\rho}^{(0)} \leftarrow$ solve (RefProb) to precision $\frac{\xi^2}{16}$ using Algorithm 8 with oracles $O_{\mathcal{C}_{\eta(\gamma-\hat{\gamma})}}$ and $O_{\mathcal{D}_{\eta\varepsilon}}$

$\tilde{\gamma}^{(0)} \leftarrow \mathrm{tr}\left(\widetilde{C}\tilde{\rho}^{(0)}\right)$

$\varepsilon_i^{(0)} \leftarrow \tilde{\rho}_{ii}^{(0)} - \frac{1}{n}$ for $i \in [n]$

$Q_{ii} \leftarrow \mathrm{sign}(-\varepsilon_i^{(0)})$ for $i \in [n]$

$\eta^{(1)} \leftarrow \frac{1}{\max\left\{\gamma - \tilde{\gamma}^{(0)}, \|\varepsilon^{(0)}\|_1\right\}}$

$\delta^{(1)} \leftarrow \frac{2}{n}\left(\|\varepsilon^{(0)}\|_1 + \frac{(\xi/4)^2}{\eta^{(1)}}\right)$

while $\max\left\{\gamma - \mathrm{tr}\left(\widetilde{C}\tilde{\rho}\right), \|\varepsilon\|_1\right\} > \zeta$ **do**

1. Store refining problem data $\left(Q \circ \widetilde{C}, \eta^{(k)}\varepsilon^{(k-1)}, \eta^{(k)}\tilde{\gamma}^{(k-1)}\right)$

2. Solve (RefProb) to precision $\frac{\xi^2}{16}$ for $\rho^{(k)}$ using Algorithm 8 with oracles $O_{\mathcal{C}_{\eta(\gamma-\hat{\gamma})}}$ and $O_{\mathcal{D}_{\eta\varepsilon}}$

3. Update solution

$$\hat{\rho}^{(k)} \leftarrow \tilde{\rho}^{(k-1)} + \frac{1}{\eta^{(k)}}Q \circ \rho^{(k)}$$

4. Apply spectrum shift to $\hat{\rho}^{(k)}$ to obtain a positive semidefinite matrix

$$\tilde{\rho}^{(k)} \leftarrow \frac{1}{1 + n\delta^{(k)}}\left(\hat{\rho}^{(k)} + \delta^{(k)}I\right)$$

5. Update objective value and compute element-wise deviations from the maximally mixed state:

$$\tilde{\gamma}^{(k)} \leftarrow \mathrm{tr}\left(\widetilde{C}\tilde{\rho}^{(k)}\right), \quad \varepsilon_i^{(k)} \leftarrow \tilde{\rho}_{ii}^{(k)} - \frac{1}{n} \text{ for } i \in [n]$$

6. Update refining problem parameters:

$$Q_{ii} \leftarrow \mathrm{sign}\left(-\varepsilon_i^{(k)}\right) \text{ for } i \in [n], \quad \eta^{(k+1)} \leftarrow \frac{1}{\max\left\{\gamma - \tilde{\gamma}^{(k)}, \|\varepsilon^{(k)}\|_1\right\}}$$

7. Update spectrum shift parameter:

$$\delta^{(k+1)} \leftarrow \frac{2}{n}\left(\|\varepsilon^{(k)}\|_1 + \frac{(\xi/4)^2}{\eta^{(k+1)}}\right)$$

8. $k \leftarrow k + 1$

end

(b) For $k \geq 1$, $\hat{\rho}^{(k)} = \tilde{\rho}^{(k-1)} + \frac{1}{\eta^{(k)}} Q \circ \rho^{(k)}$ satisfies

$$\max\left\{ \gamma - \mathrm{tr}\left(\widetilde{C}\hat{\rho}^{(k)}\right), \left\|\sum_{i\in[n]} \langle i|\hat{\rho}^{(k)}|i\rangle |i\rangle\langle i| - n^{-1}I\right\|_{\mathrm{tr}} \right\} \leq \xi^{k+2}, \quad \lambda_{\min}\left(\hat{\rho}^{(k)}\right) \geq -\frac{\xi^{k+1}}{n}.$$

(c) For $k \geq 0$, $\tilde{\rho}^{(k)}$ satisfies

$$\max\left\{ \gamma - \mathrm{tr}\left(\widetilde{C}\tilde{\rho}^{(k)}\right), \left\|\sum_{i\in[n]} \langle i|\tilde{\rho}^{(k)}|i\rangle |i\rangle\langle i| - n^{-1}I\right\|_{\mathrm{tr}} \right\} \leq 2\xi^{k+1}, \quad \lambda_{\min}\left(\tilde{\rho}^{(k)}\right) \geq 0.$$

That is, $\tilde{\rho}^{(k)}$ is an $2\xi^{k+1}$-precise solution to (OptRelaxed).

Proof. First, observe that that we initialize $\varepsilon_i = \frac{1}{n}$ for $i \in [n]$, $\tilde{\gamma} = 0$, $\eta^{(0)} = 1$ and $Q = ee^{\top}$. Under these conditions, one can observe that if $\tilde{\rho}^{(0)}$ is obtained from solving (RefProb) to precision $\tilde{\xi}^2$ using the oracles $O_{C_{\eta(\gamma-\tilde{\gamma})}}$ and $O_{D_{\eta\varepsilon}}$, we must have

$$\sum_{i=1}^{n} \left| \left\langle i \left| \tilde{\rho}^{(0)} \right| i \right\rangle - \frac{1}{n} \right| \leq \frac{\tilde{\xi}^2}{\eta^{(0)}} = \tilde{\xi}^2. \tag{7.12}$$

In other words, $\tilde{\rho}^{(0)}$ satisfies

$$\left\|\sum_{i\in[n]} \langle i|\tilde{\rho}^{(0)}|i\rangle |i\rangle\langle i| - n^{-1}I\right\|_{\mathrm{tr}} \leq \tilde{\xi}^2,$$

and by the definition of $O_{C_{\eta(\gamma-\tilde{\gamma})}}$ we also have

$$\mathrm{tr}\left(\widetilde{C}\tilde{\rho}^{(0)}\right) \geq \gamma - \frac{\tilde{\xi}^2}{\eta^{(0)}} = \gamma - \tilde{\xi}^2. \tag{7.13}$$

Since $\tilde{\rho}^{(0)} \succeq 0$ by construction, clearly $\tilde{\rho}^{(0)}$ is a $\tilde{\xi}^2$-precise solution to (OptRelaxed).

Next, we proceed by induction to establish that for $k \geq 1$, the matrix $\hat{\rho}^{(k)}$ satisfies

$$\max\left\{ \gamma - \mathrm{tr}\left(\widetilde{C}\hat{\rho}^{(k)}\right), \left\|\sum_{i\in[n]} \langle i|\hat{\rho}^{(k)}|i\rangle |i\rangle\langle i| - n^{-1}I\right\|_{\mathrm{tr}} \right\} \leq \frac{\tilde{\xi}^2}{\eta^{(k)}},$$

$$\lambda_{\min}\left(\hat{\rho}^{(k)}\right) \geq -\frac{2}{n}\left(\frac{\tilde{\xi}^2}{\eta^{(k-1)}} + \frac{\tilde{\xi}^2}{\eta^{(k)}}\right). \tag{7.14}$$

For all $k \geq 1$, we have $\varepsilon_i^{(k-1)} = \tilde{\rho}_{ii}^{(k-1)} - \frac{1}{n}$ for $i \in [n]$ and $Q = (ee^{\top} - I) + \mathrm{diag}\left(\mathrm{sign}\left(-\varepsilon^{(k-1)}\right)\right)$. For this choice of parameters, the general feasibility problem (RefProb) reduces to the refining problem (7.9) and the solution $\rho^{(k)}$ obtained via Algorithm 8 using the oracles $O_{C_{\eta(\gamma-\tilde{\gamma})}}$ and

$O_{\mathcal{D}_{\eta\varepsilon}}$ must satisfy

$$\operatorname{tr}\left(\widetilde{C}Q\circ\rho^{(k)}\right) \geq \eta^{(k)}\left(\gamma - \tilde{\gamma}^{(k-1)}\right) - \tilde{\xi}^2 \tag{7.15a}$$

$$\sum_{i=1}^{n}\left|\left\langle i\left|\rho^{(k)}\right|i\right\rangle - \eta^{(k)}|\varepsilon_i^{(k-1)}|\right| \leq \tilde{\xi}^2. \tag{7.15b}$$

Accordingly, for $k \geq 1$, setting $\hat{\rho}^{(k)} = \tilde{\rho}^{(k-1)} + \frac{1}{\eta^{(k)}}Q\circ\rho^{(k)}$ we can bound the total infeasibility of the diagonal constraints as follows

$$
\begin{aligned}
\sum_{i=1}^{n}\left|\left\langle i\left|\hat{\rho}^{(k)}\right|i\right\rangle - \frac{1}{n}\right| &= \sum_{i=1}^{n}\left|\left\langle i\left|\tilde{\rho}^{(k-1)} + \frac{1}{\eta^{(k)}}Q\circ\rho^{(k)}\right|i\right\rangle - \frac{1}{n}\right| \\
&= \sum_{i=1}^{n}\left|\left(\tilde{\rho}_{ii}^{(k-1)} + \frac{1}{\eta^{(k)}}\left(\operatorname{sign}\left(-\varepsilon_i^{(k-1)}\right)\cdot\rho_{ii}^{(k)}\right)\right) - \frac{1}{n}\right| \\
&= \sum_{i=1}^{n}\left|\left(\tilde{\rho}_{ii}^{(k-1)} - \frac{1}{n}\right) + \frac{1}{\eta^{(k)}}\operatorname{sign}\left(-\varepsilon_i^{(k-1)}\right)\rho_{ii}^{(k)}\right| \\
&= \sum_{i=1}^{n}\left|\varepsilon_i^{(k-1)} + \frac{1}{\eta^{(k)}}\operatorname{sign}\left(-\varepsilon_i^{(k-1)}\right)\rho_{ii}^{(k)}\right| \\
&= \frac{1}{\eta^{(k)}}\sum_{i=1}^{n}\left|\eta^{(k)}\varepsilon_i^{(k-1)} + \operatorname{sign}\left(-\varepsilon_i^{(k-1)}\right)\rho_{ii}^{(k)}\right| \leq \frac{\tilde{\xi}^2}{\eta^{(k)}},
\end{aligned}
$$

where the final inequality follows from (7.15b). Consequently, we can conclude that at iteration $k \geq 1$ we have

$$\left\|\sum_{i\in[n]}\langle i|\hat{\rho}^{(k)}|i\rangle|i\rangle\langle i| - n^{-1}I\right\|_{\operatorname{tr}} \leq \frac{\tilde{\xi}^2}{\eta^{(k)}}. \tag{7.16}$$

Next, letting $\tilde{c}^{(k)} = \operatorname{tr}\left(\widetilde{C}Q\circ\rho^{(k)}\right)$, one can observe

$$\operatorname{tr}\left(\widetilde{C}\hat{\rho}^{(k)}\right) = \operatorname{tr}\left(\widetilde{C}\left(\tilde{\rho}^{(k-1)} + \frac{1}{\eta^{(k)}}Q\circ\rho^{(k)}\right)\right) = \operatorname{tr}\left(\widetilde{C}\tilde{\rho}^{(k-1)}\right) + \frac{1}{\eta^{(k)}}\operatorname{tr}\left(\widetilde{C}Q\circ\rho^{(k)}\right) = \tilde{\gamma}^{(k-1)} + \frac{\tilde{c}^{(k)}}{\eta^{(k)}}.$$

Since $\tilde{c}^{(k)} \geq \eta^{(k)}\left(\gamma - \tilde{\gamma}^{(k-1)}\right) - \xi^2$ due to (7.15a), it follows:

$$\operatorname{tr}\left(\widetilde{C}\hat{\rho}^{(k)}\right) = \tilde{\gamma}^{(k-1)} + \frac{\tilde{c}^{(k)}}{\eta^{(k)}} \geq \tilde{\gamma}^{(k-1)} + \frac{1}{\eta^{(k)}}\left[\eta^{(k)}\left(\gamma - \tilde{\gamma}^{(k-1)}\right) - \tilde{\xi}^2\right] = \gamma - \frac{\tilde{\xi}^2}{\eta^{(k)}}. \tag{7.17}$$

To prove the eigenvalue bound on $\hat{\rho}^{(k)}$, we first establish that $\tilde{\rho}^{(k)} \succeq 0$ holds for all $k \geq 0$ using induction. When $k = 0$, this is trivially true because $\tilde{\rho}^{(0)} \succ 0$ by construction (it is a Gibbs state). Now assume that $\tilde{\rho}^{(\ell)} \succeq 0$ holds for all $\ell = 1,\ldots,k-1$. At the k-th iterate, we have

$$\tilde{\rho}^{(k)} := \frac{1}{1+n\delta^{(k)}}\left(\hat{\rho}^{(k)} + \delta^{(k)}I\right) = \frac{1}{1+n\delta^{(k)}}\left[\tilde{\rho}^{(k-1)} + \frac{1}{\eta^{(k)}}Q\circ\rho^{(k)} + \delta^{(k)}I\right].$$

Since we define $\delta^{(k)} = \frac{2}{n}\left(\left\|\varepsilon^{(k-1)}\right\|_1 + \frac{\tilde{\xi}^2}{\eta^{(k)}}\right)$, Proposition 7.15 asserts that $\frac{1}{\eta^{(k)}}Q\circ\rho^{(k)} + \delta^{(k)}I \succeq 0$.

Combining this fact with $\tilde{\rho}^{(k-1)} \succeq 0$, we have $\tilde{\rho}^{(k)} \succeq 0$ because it is defined as the sum of two symmetric positive semidefinite matrices, thus completing the induction argument.

Having shown $\tilde{\rho}^{(k)} \succeq 0$ holds for all $k \geq 0$, it follows that $\hat{\rho}^{(k)}$ is defined as the sum of a positive semidefinite matrix, and a symmetric matrix satisfying $\frac{1}{\eta^{(k)}} Q \circ \rho^{(k)} \succeq -\delta^{(k)} I$. To see this, first observe that the residuals along the diagonal ε are always computed with respect to $\tilde{\rho}$, and in particular for $k \geq 1$

$$
\begin{aligned}
\left\| \varepsilon^{(k)} \right\|_1 &= \left\| \sum_{i \in [n]} \langle i | \tilde{\rho}^{(k)} | i \rangle | i \rangle\langle i | - n^{-1} I \right\|_{\mathrm{tr}} = \sum_{i \in [n]} \left| \frac{1}{1 + n\delta^{(k)}} \left(\hat{\rho}_{ii}^{(k)} + \delta^{(k)} \right) - n^{-1} \right| \\
&= \frac{1}{1 + n\delta^{(k)}} \sum_{i \in [n]} \left| \hat{\rho}_{ii}^{(k)} + \delta^{(k)} - n^{-1} \left(1 + n\delta^{(k)} \right) \right| \\
&= \frac{1}{1 + n\delta^{(k)}} \sum_{i \in [n]} \left| \hat{\rho}_{ii}^{(k)} + \delta^{(k)} - n^{-1} - \delta^{(k)} \right| \\
&\leq \frac{1}{1 + n\delta^{(k)}} \sum_{i \in [n]} \left| \hat{\rho}_{ii}^{(k)} - n^{-1} \right| \\
&\leq \frac{1}{1 + n\delta^{(k)}} \cdot \frac{\tilde{\xi}^2}{\eta^{(k)}},
\end{aligned}
\tag{7.18}
$$

where the final inequality follows from (7.16). Thus, applying Lemma 7.10 along with the definition of $\delta^{(k)}$ and noting (7.12), (7.18) we obtain

$$
\delta^{(k)} = \frac{2}{n} \left(\left\| \varepsilon^{(k-1)} \right\|_1 + \frac{\tilde{\xi}^2}{\eta^{(k)}} \right) \leq \frac{2}{n} \left(\frac{\tilde{\xi}^2}{\eta^{(k-1)}} + \frac{\tilde{\xi}^2}{\eta^{(k)}} \right) \implies \hat{\rho}^{(k)} \succeq -2 \left(\frac{\tilde{\xi}^2}{\eta^{(k-1)}} + \frac{\tilde{\xi}^2}{\eta^{(k)}} \right) n^{-1} I,
\tag{7.19}
$$

for all $k \geq 1$. Hence, from (7.16), (7.17) and (7.19), the matrix $\hat{\rho}^{(k)} = \tilde{\rho}^{(k-1)} + \frac{1}{\eta^{(k)}} Q \circ \rho^{(k)}$ satisfies (7.14) for all $k \geq 0$.

Next, note that the spectrum shift used to restore positive semidefiniteness is mild. Indeed for all $k \geq 1$

$$
\begin{aligned}
\left\| \hat{\rho}^{(k)} - \tilde{\rho}^{(k)} \right\|_{\mathrm{tr}} &= \left\| \hat{\rho}^{(k)} - \frac{1}{1 + n\delta^{(k)}} \left(\hat{\rho}^{(k)} + \delta^{(k)} I \right) \right\|_{\mathrm{tr}} \\
&= \left\| \frac{1 + n\delta^{(k)} - 1}{1 + n\delta^{(k)}} \hat{\rho}^{(k)} - \frac{1}{1 + n\delta^{(k)}} \delta^{(k)} I \right\|_{\mathrm{tr}} \\
&\leq \frac{n\delta^{(k)}}{1 + n\delta^{(k)}} \left\| \hat{\rho}^{(k)} - n^{-1} I \right\|_{\mathrm{tr}} \\
&= \frac{n\delta^{(k)}}{1 + n\delta^{(k)}} \left\| \hat{\rho}^{(k)} - n^{-1} I + (\delta^{(k)} I - \delta^{(k)} I) \right\|_{\mathrm{tr}} \\
&\leq \frac{n\delta^{(k)}}{1 + n\delta^{(k)}} \left[\left\| \hat{\rho}^{(k)} + \delta^{(k)} I \right\|_{\mathrm{tr}} + \left\| n^{-1} I \right\|_{\mathrm{tr}} + \left\| \delta^{(k)} I \right\|_{\mathrm{tr}} \right] \\
&\leq \frac{n\delta^{(k)}}{1 + n\delta^{(k)}} \left[1 + \frac{\tilde{\xi}^2}{\eta^{(k)}} + n\delta^{(k)} + 1 + n\delta^{(k)} \right] \\
&= \left(2 + \frac{\tilde{\xi}^2}{\eta^{(k)}} + 2n\delta^{(k)} \right) \frac{n\delta^{(k)}}{1 + n\delta^{(k)}},
\end{aligned}
$$

where the second to last inequality follows from the fact that $\hat{\rho}^{(k)} + \delta^{(k)} I$, $n^{-1} I$ and $\delta^{(k)} I$ are all positive semidefinite matrices.

Using the definition of $\delta^{(k)}$, we can continue the chain of inequalities:

$$
\left(2 + \frac{\tilde{\xi}^2}{\eta^{(k)}} + 2n\delta^{(k)} \right) \frac{n\delta^{(k)}}{1 + n\delta^{(k)}} < \left(2 + \frac{\tilde{\xi}^2}{\eta^{(k)}} + 2n\delta^{(k)} \right) n\delta^{(k)}
$$

$$
\leq 2 \left(2 + 3\frac{\tilde{\xi}^2}{\eta^{(k)}} + 2 \left\| \varepsilon^{(k-1)} \right\|_1 \right) \left(\left\| \varepsilon^{(k-1)} \right\|_1 + \frac{\tilde{\xi}^2}{\eta^{(k)}} \right)
$$

$$
= 4 \left[\left\| \varepsilon^{(k-1)} \right\|_1 + \frac{5}{2} \frac{\tilde{\xi}^2}{\eta^{(k)}} \left\| \varepsilon^{(k-1)} \right\|_1 + \frac{3}{2} \frac{\tilde{\xi}^4}{\left(\eta^{(k)}\right)^2} + \left\| \varepsilon^{(k-1)} \right\|_1^2 \right]
$$

$$
\leq 4 \left[\frac{\tilde{\xi}^2}{\eta^{(k-1)}} + \frac{5}{2} \frac{\tilde{\xi}^2}{\eta^{(k-1)}} \frac{\tilde{\xi}^2}{\eta^{(k)}} + \frac{3}{2} \frac{\tilde{\xi}^4}{\left(\eta^{(k)}\right)^2} + \frac{\tilde{\xi}^4}{\left(\eta^{(k-1)}\right)^2} \right],
$$

where the final inequality follows upon noting (7.12) and (7.18). For ease of notation, for all $k \geq 1$ define

$$
\Phi\left(\eta^{(k-1)}, \eta^{(k)}, \tilde{\xi} \right) := 4 \left[\frac{\tilde{\xi}^2}{\eta^{(k-1)}} + \frac{5}{2} \frac{\tilde{\xi}^2}{\eta^{(k-1)}} \frac{\tilde{\xi}^2}{\eta^{(k)}} + \frac{3}{2} \frac{\tilde{\xi}^4}{\left(\eta^{(k)}\right)^2} + \frac{\tilde{\xi}^4}{\left(\eta^{(k-1)}\right)^2} \right].
$$

Applying a matrix Hölder inequality, one can observe that for all $k \geq 1$:

$$
\left| \mathrm{tr}\left(\tilde{C} \hat{\rho}^{(k)} \right) - \mathrm{tr}\left(\tilde{C} \bar{\rho}^{(k)} \right) \right| \leq \left\| \tilde{C} \right\| \left\| \hat{\rho}^{(k)} - \bar{\rho}^{(k)} \right\|_{\mathrm{tr}} \leq \left\| \hat{\rho}^{(k)} - \bar{\rho}^{(k)} \right\|_{\mathrm{tr}} < \Phi\left(\eta^{(k-1)}, \eta^{(k)}, \tilde{\xi} \right),
$$

from which we can conclude

$$
\gamma - \mathrm{tr}\left(\tilde{C} \bar{\rho}^{(k)} \right) = \gamma - \mathrm{tr}\left(\tilde{C} \rho^{(k)} \right) + \left[\mathrm{tr}\left(\tilde{C} \rho^{(k)} \right) - \mathrm{tr}\left(\tilde{C} \hat{\rho}^{(k)} \right) \right]
$$

$$
= \gamma - \mathrm{tr}\left(\tilde{C} \hat{\rho}^{(k)} \right) + \left[\mathrm{tr}\left(\tilde{C} \hat{\rho}^{(k)} \right) - \mathrm{tr}\left(\tilde{C} \bar{\rho}^{(k)} \right) \right]
$$

$$
\leq \frac{\tilde{\xi}^2}{\eta^{(k)}} + \Phi\left(\eta^{(k-1)}, \eta^{(k)}, \tilde{\xi} \right), \tag{7.20}
$$

holds for all $k \geq 1$, as $\gamma - \mathrm{tr}\left(\tilde{C} \hat{\rho}^{(k)} \right) \leq \frac{\tilde{\xi}^2}{\eta^{(k)}}$. We can now use this fact to establish the lower bound on $\eta^{(k)}$, which we prove by induction.

For $k = 0$, we have $\eta^{(0)} = 1$ and $\eta^{(1)} = \frac{1}{\tilde{\xi}^2}$ when $k = 1$ due to (7.12) and (7.13). for which $\eta^{(k)} \geq \frac{1}{2 \cdot \tilde{\xi}^k}$ trivially holds. By the induction hypothesis, it assumed that $\eta^{(\ell)} \geq \frac{1}{2\tilde{\xi}^\ell}$ is true for

$\ell = 2, \ldots, k$. From here, one can observe that

$$\Phi\left(\eta^{(k-1)}, \eta^{(k)}, \tilde{\xi}\right) = 4\left[\frac{\tilde{\xi}^2}{\eta^{(k-1)}} + \frac{5}{2}\frac{\tilde{\xi}^2}{\eta^{(k-1)}}\frac{\tilde{\xi}^2}{\eta^{(k)}} + \frac{3}{2}\frac{\tilde{\xi}^4}{\left(\eta^{(k)}\right)^2} + \frac{\tilde{\xi}^4}{\left(\eta^{(k-1)}\right)^2}\right]$$

$$\leq 4\left[2\tilde{\xi}^2\xi^{k-1} + \frac{5}{2}\tilde{\xi}^4\xi^{k-1}\xi^k + 6\tilde{\xi}^4\xi^{2k} + 4\tilde{\xi}^4\xi^{2[k-1]}\right]$$

$$= 8\tilde{\xi}^2\xi^{k-1} + 10\tilde{\xi}^4\xi^{k-1}\xi^k + 24\tilde{\xi}^4\xi^{2k} + 16\tilde{\xi}^4\xi^{2[k-1]}$$

$$\leq \frac{8}{16}\xi^{k+1} + \frac{10}{256}\xi^{2k+3} + \frac{24}{64}\xi^{2k+4} + \frac{16}{64}\xi^{2k+2}$$

$$< \frac{178}{256}\xi^{k+1}.$$

Noting $\Phi\left(\eta^{(k-1)}, \eta^{(k)}, \tilde{\xi}\right) < \frac{178}{256}\xi^{k+1}$ and applying (7.14) yields

$$\eta^{(k+1)} = \frac{1}{\max\left\{\gamma - \mathrm{tr}\left(\widetilde{C}\tilde{\rho}^{(k)}\right), \left\|\sum_{i\in[n]}\langle i|\tilde{\rho}^{(k)}|i\rangle|i\rangle\langle i| - n^{-1}I\right\|_{\mathrm{tr}}\right\}} \geq \frac{1}{\frac{\tilde{\xi}^2}{\eta^{(k)}} + \Phi\left(\eta^{(k-1)}, \eta^{(k)}, \tilde{\xi}\right)}$$

$$\geq \frac{1}{\frac{\tilde{\xi}^2}{\eta^{(k)}} + \frac{178}{256}\xi^{k+1}}$$

$$\geq \frac{1}{2\tilde{\xi}^2\xi^k + \frac{178}{256}\xi^{k+1}} > \frac{1}{2\xi^{k+1}},$$

which completes the proof of *(a)*.

Having demonstrated that *(a)* holds, to prove *(b)*, we can simply combine inequality (7.14) with the lower bound $\eta^{(k)} \geq \frac{1}{2\xi^k}$, which together imply

$$\max\left\{\gamma - \mathrm{tr}\left(\widetilde{C}\hat{\rho}^{(k)}\right), \left\|\sum_{i\in[n]}\langle i|\hat{\rho}^{(k)}|i\rangle|i\rangle\langle i| - n^{-1}I\right\|_{\mathrm{tr}}\right\} \leq \frac{\tilde{\xi}^2}{\eta^{(k)}} \leq \xi^{k+2}, \quad \lambda_{\min}\left(\hat{\rho}^{(k)}\right) \geq -\frac{\xi^{k+1}}{n}.$$

Upon noting (7.18), (7.20) and that $\tilde{\rho}^{(k)} \succeq 0$ always holds, the result in *(c)* follows from a similar argument. $\qquad\square$

The next result establishes polynomial convergence of Algorithm 9.

Corollary 7.4. *Let $0 < \zeta \ll \xi < 1$, and $\eta^{(0)} = 1$. Then, Algorithm 9 terminates in at most*

$$K = \mathcal{O}\left(\log\left(\frac{1}{\zeta}\right)\right)$$

iterations.

Proof. The result follows from Theorem 7.16*(c)*. $\qquad\square$

It is important at this point for us to remark that fixing $\xi \in (0,1)$ does not limit us with respect to how accurately we can solve (QUBO-SDO). We can always make the final precision parameter arbitrarily small using only $\widetilde{\mathcal{O}}_{\frac{1}{\xi}}(1)$ iterations, as the overall running time depends only poly-logarithmically on ζ^{-1}. Accordingly, we take advantage of this fact and revisit the approximation guarantee provided in Proposition 7.6.

Proposition 7.17. *Let $\tilde{\rho}$ be a ζ-accurate solution to the renormalized and relaxed SDO problem (OptRelaxed) with input matrix C and $\zeta = \left(\frac{\epsilon}{n\|C\|_F}\right)^4$. Let $\gamma_\zeta = \mathrm{tr}\,(C\tilde{\rho})$ be the value attained by $\tilde{\rho}$. Then, there is a quantum state ρ^* at trace distance $\mathcal{O}\left(\frac{\epsilon}{n\|C\|_F}\right)$ of $\tilde{\rho}$ such that $n\rho^*$ is a feasible point of SDO problem (QUBO-SDO). In particular*

$$\left|\gamma_\zeta n\|C\|_F - \mathrm{tr}\,(n\rho^* C)\right| = \mathcal{O}\,(\epsilon).$$

Moreover, it is possible to construct ρ^* in time $\mathcal{O}(n^2)$ given the entries of $\tilde{\rho}$.

Proof. The proof almost exactly follows the proof of Proposition 3.1 in [BKF22], regardless, we present the adjusted proof for completeness. Our aim is to show that a ζ-precise solution $\tilde{\rho}$ to (OptRelaxed) obtained using Algorithm 9 can be used to construct ρ^* such that $n\rho^*$ is an exactly feasible solution to (QUBO-SDO).

We begin by examining the diagonal elements of $\tilde{\rho}$ and check whether modifications need to be made to ensure that our solution is an exactly feasible point to the renormalized SDO problem (OptRelaxed). Namely, if $|\langle i|\tilde{\rho}|i\rangle - \frac{1}{n}| > \frac{\sqrt{\zeta}}{n}$ for $i \in [n]$, we replace $\tilde{\rho}_{ii}$ with $\frac{1}{n}$ and set all elements in the i-th row and the i-th column to 0, and denote the resulting matrix by ρ'. From here we introduce another matrix W which we obtain by replacing each diagonal entry of ρ' with $\frac{1}{n}$. In general we may not have $W \succeq 0$, so the authors in [BKF22] suggest using the convex combination:

$$\rho^* = \frac{1}{1 + \sqrt{\zeta}}\left(W + \frac{\sqrt{\zeta}}{n}I\right).$$

Then, $\rho^* \succeq 0$ and by construction $\langle i|\rho^*|i\rangle = \frac{1}{n}$ for all $i \in [n]$. Hence, ρ^* is a feasible solution to the renormalized SDO problem (OptRelaxed).

What remains is to show that the above reformulations yield the desired approximation. Denote by $\mathcal{B} = \{i : |n\langle i|\tilde{\rho}|i\rangle - 1| > \sqrt{\zeta}\} \subset [n]$ the set of diagonal entries that deviate substantially from $\frac{1}{n}$. Without loss of generality, it suffices to assume that such elements are found in the first $|\mathcal{B}|$ rows of $\tilde{\rho}$, in which case

$$\|\rho' - \tilde{\rho}\|_{\mathrm{tr}} = \left\|\begin{pmatrix} n^{-1}I_\mathcal{B} & 0 \\ 0 & \tilde{\rho}_{22} \end{pmatrix} - \begin{pmatrix} \tilde{\rho}_{11} & \tilde{\rho}_{12} \\ \tilde{\rho}_{21} & \tilde{\rho}_{22} \end{pmatrix}\right\|_{\mathrm{tr}} = \left\|\begin{pmatrix} n^{-1}I_\mathcal{B} - \tilde{\rho}_{11} & -\tilde{\rho}_{12} \\ -\tilde{\rho}_{21} & 0 \end{pmatrix}\right\|_{\mathrm{tr}}$$
$$\leq \|\tilde{\rho}_{11}\|_{\mathrm{tr}} + 2\|\tilde{\rho}_{12}\|_{\mathrm{tr}} + \|n^{-1}I_\mathcal{B}\|_{\mathrm{tr}}. \tag{7.21}$$

Since $\tilde{\rho}$ is a ζ-precise solution to (OptRelaxed), $\tilde{\rho}$ obeys

$$\sum_{i=1}^{n}\left|\langle i|\tilde{\rho}|i\rangle - \frac{1}{n}\right| \leq \zeta.$$

Therefore, we must have

$$|\mathcal{B}|\frac{\sqrt{\zeta}}{n} \le \zeta,$$

which equates to $|\mathcal{B}| \le n\sqrt{\zeta}$. Now, by the definition of $\mathcal{B}$, it follows

$$\|\tilde{\rho}_{22}\|_{\mathrm{tr}} \ge (n - |\mathcal{B}|)\frac{1 - \sqrt{\zeta}}{n} \ge (n - n\sqrt{\zeta})\frac{1 - \sqrt{\zeta}}{n} = (1 - \sqrt{\zeta})^2.$$

Following [BKF22], we invoke a result from [Kin03], which states

$$\left\|\begin{bmatrix} \|\tilde{\rho}_{11}\|_{\mathrm{tr}} & \|\tilde{\rho}_{12}\|_{\mathrm{tr}} \\ \|\tilde{\rho}_{12}^\top\|_{\mathrm{tr}} & \|\tilde{\rho}_{22}\|_{\mathrm{tr}} \end{bmatrix}\right\|_{\mathrm{tr}} \le \left\|\begin{bmatrix} \tilde{\rho}_{11} & \tilde{\rho}_{12} \\ \tilde{\rho}_{12}^\top & \tilde{\rho}_{22} \end{bmatrix}\right\|_{\mathrm{tr}} = \|\tilde{\rho}\|_{\mathrm{tr}} = \mathrm{tr}\,(\tilde{\rho}) = 1.$$

Using the fact that $\|\cdot\|_{\mathrm{tr}} \ge \|\cdot\|_2$, where $\|\cdot\|_2$ is the Frobenius, or Schatten-2 norm, the above implies

$$\|\tilde{\rho}_{11}\|_{\mathrm{tr}}^2 + 2\|\tilde{\rho}_{12}\|_{\mathrm{tr}}^2 + \|\tilde{\rho}_{22}\|_{\mathrm{tr}}^2 \le 1.$$

As $\|\tilde{\rho}_{22}\|_{\mathrm{tr}} \ge (1 - \sqrt{\zeta})^2$, it can be seen trivially that $\|\tilde{\rho}_{22}\|_{\mathrm{tr}}^2 \ge (1 - \sqrt{\zeta})^4$, and thus

$$\|\tilde{\rho}_{11}\|_{\mathrm{tr}}^2 + 2\|\tilde{\rho}_{12}\|_{\mathrm{tr}}^2 \le 1 - (1 - \sqrt{\zeta})^4 = \mathcal{O}(\sqrt{\zeta}).$$

Consequently $\|\tilde{\rho}_{11}\|_{\mathrm{tr}} + 2\|\tilde{\rho}_{12}\|_{\mathrm{tr}} = \mathcal{O}\left(\zeta^{\frac{1}{4}}\right)$, and plugging this into equation (7.21) asserts

$$\|\rho' - \tilde{\rho}\|_{\mathrm{tr}} = \mathcal{O}\left(\zeta^{\frac{1}{4}}\right). \tag{7.22}$$

Let R be a diagonal matrix whose elements are $R_{ii} \in \left[-\frac{\sqrt{\zeta}}{n}, \frac{\sqrt{\zeta}}{n}\right]$ for $i \in [n]$, such that

$$W = \rho' + R,$$

and note that $R + \sqrt{\zeta}n^{-1}I \succeq 0$. Upon normalizing the trace, one can observe

$$\rho^* = \frac{1}{1 + \sqrt{\zeta}}\left(\rho' + R + \sqrt{\zeta}n^{-1}I\right) \succeq 0,$$

with $\rho_{ii}^* = \frac{1}{n}n$ for all $i \in [n]$. Thus, $n\rho^*$ is a feasible solution to the SDO problem (QUBO-SDO). Further, by a triangle inequality we have

$$\|\rho' - \rho^*\|_{\mathrm{tr}} = \frac{1}{1 + \sqrt{\zeta}}\left\|\sqrt{\zeta}\rho' + R + \sqrt{\zeta}n^{-1}I\right\|_{\mathrm{tr}} = \mathcal{O}(\sqrt{\zeta}). \tag{7.23}$$

Combining equations (7.22) and (7.23) and noting $\zeta = \left(\frac{\epsilon}{n\|C\|_F}\right)^4$, applying another triangle inequality yields

$$\|\tilde{\rho} - \rho^*\|_{\mathrm{tr}} = \mathcal{O}\left(\zeta^{\frac{1}{4}}\right) = \mathcal{O}\left(\left[\left(\frac{\epsilon}{n\|C\|_F}\right)^4\right]^{\frac{1}{4}}\right) = \mathcal{O}\left(\frac{\epsilon}{n\|C\|_F}\right).$$

Then, the result follows from a matrix Hölder inequality:

$$\left| \operatorname{tr}(nC\rho) - \operatorname{tr}(nC\rho^*) \right| \leq n\|C\|\|\rho - \rho^*\|_{\mathrm{tr}} = \mathcal{O}\left(n\|C\|_F \zeta^{\frac{1}{4}}\right) = \mathcal{O}\left(n\|C\|_F \left[\frac{\epsilon}{n\|C\|_F}\right]\right) = \mathcal{O}(\epsilon).$$

$\square$

7.5 Complexity

We now analyze the worst case overall running time of our Iterative Refinement Method given in Algorithm 9 in both the classical and quantum settings.

7.5.1 Classical running time

As we saw in Section 7.3, the complexity of using Algorithm 8 to solve the SDO problem (QUBO-SDO) scales poorly in the inverse precision, with the classical algorithm exhibiting an $\mathcal{O}(\epsilon^{-12})$ dependence. In both the classical and quantum cases, our iterative refinement scheme reconciles the poor scaling in ϵ because it possesses the following two properties. First, we can obtain an arbitrarily precise solution to (OptRelaxed) in at most $\tilde{\mathcal{O}}_{\frac{1}{\xi}}(1)$ iterations. Second, it suffices to treat ξ as fixed for the oracle calls that occur in each iteration, as the precision of the final solution is a byproduct of how we use these solution of the refining problems to produce a solution to (OptRelaxed).

The next result formalizes the above argument, and establishes the complexity of Algorithm 9 for the classical case.

Theorem 7.18. *Let* $C \in \mathcal{S}^n$ *with row sparsity* s *and* $\epsilon \in (0, 1)$. *Then, fixing* $\xi \in (0, 1)$ *with* $0 < \epsilon \ll \xi < 1$, *and setting* $\zeta = \left(\frac{\epsilon}{n\|C\|_F}\right)^4$, *a classical implementation of Algorithm 9 solves* (QUBO-SDO) *up to additive error* $\mathcal{O}(\epsilon)$ *in time*

$$\mathcal{O}\left(\min\{n^2 s, n^\omega\} \cdot \operatorname{polylog}\left(n, \|C\|_F, \frac{1}{\epsilon}\right)\right).$$

The output of the algorithm is a classical description of a matrix $\tilde{\rho} \in \mathcal{S}^n_+$ *that is a* ζ-*precise solution to* (OptRelaxed). *The entries of* $\tilde{\rho}$ *can be modified to construct a matrix* ρ^* *at trace distance* $\mathcal{O}\left(\frac{\epsilon}{n\|C\|_F}\right)$ *of* $\tilde{\rho}$ *in time* $\mathcal{O}(n^2)$, *such that* $n\rho^*$ *is a feasible point of the SDO problem* (QUBO-SDO).

Proof. Given that C is an s-sparse matrix, we can load C in $\mathcal{O}(ns)$ time, and from here we must compute $\|C\|_F$, which requires $\mathcal{O}(ns)$ arithmetic operations. In every iteration of Algorithm 9, we make a call to our subroutine in Algorithm 8, before updating the solution and preparing the next refining problem. Updating the solution involves matrix addition between two $n \times n$ matrices and requires $\mathcal{O}(n^2)$ arithmetic operations, whereas updating Q and ε for the next refining problem can be accomplished using $\mathcal{O}(n)$ arithmetic operations, as only the diagonal entries of Q need to be stored and maintained.

In view of Proposition 7.7, the dominant operation at each iteration is the use of Algorithm 8 to solve the SDO problem at hand. By Proposition 7.7, Algorithm 8 can be used to solve (RefProb) to additive error ξ in time

$$\mathcal{T}_{HU}^{\text{classical}} = \mathcal{O}\left(\min\{n^2 s, n^\omega\} \log^2(n) \xi^{-3}\right).$$

If every call to Algorithm 8 is made using precision ξ^2, then by Corollary 7.4, Algorithm 9 converges in at most $\mathcal{O}\left(\log(\zeta^{-1})\right)$ iterations, and we can thus express the overall running time of Algorithm 9 as

$$\mathcal{O}\left(\left(\min\{n^2 s, n^\omega\} \log^2(n) \xi^{-6}\right) \log(\zeta^{-1})\right).$$

In the context of Algorithm 9, it suffices to carry out each of the calls to the SDO subroutine (calls to Algorithm 8) using fixed (i.e., constant) precision ξ to obtain a ζ-precise solution to (OptRelaxed). The above complexity thus reduces to

$$\mathcal{O}\left(\min\{n^2 s, n^\omega\} \log^2(n) \log(\zeta^{-1})\right).$$

For our choice of $\zeta = \left(\frac{\epsilon}{n\|C\|_F}\right)^4$, one can observe

$$\mathcal{O}\left(\min\{n^2 s, n^\omega\} \log^2(n) \log(\zeta^{-1})\right) = \mathcal{O}\left(\min\{n^2 s, n^\omega\} \cdot \text{polylog}\left(n, \|C\|_F, \frac{1}{\epsilon}\right)\right).$$

Proposition 7.17 certifies that the above running time suffices to obtain a ρ from which we can construct ρ^* in time $\mathcal{O}(n^2)$, such that $n\rho^*$ is a feasible point of the SDO problem (QUBO-SDO) satisfying

$$\left|\gamma_\zeta n\|C\|_F - \text{tr}\left(n\rho^* C\right)\right| = \mathcal{O}(\epsilon),$$

and the proof is complete. $\qquad\qquad\qquad\qquad\qquad\qquad\qquad\qquad\qquad\qquad\square$

7.5.2 Quantum running time

Just as in the classical case, we show that a quantum implementation of Algorithm 9 mitigates the poor scaling in the running time with respect to the inverse precision.

Our quantum implementation of Algorithm 9 is provided in Algorithm 10. The relevant error parameters are the same as those appearing in Algorithm 9: *(i)* ξ, the fixed precision used to test closeness to the sets $\mathcal{C}_{\eta(\gamma-\bar{\gamma})}$ and $\mathcal{D}_{\eta\varepsilon}$ in every iteration, *(ii)* ζ, the precision to which the final solution solves (OptRelaxed), and *(iii)* ϵ, the additive error to which we seek to solve (QUBO-SDO). In our initialization steps we set the values of Q, ε and η such that the first call to Algorithm 8 solves the original feasibility problem (7.8). We also create a vector $p = 0^{n\times 1}$ that will be used to maintain a classical description of the diagonal elements of our solution over the course of the algorithm.

At every iteration k, a call is made to Algorithm 8 with separation oracles $O_{\mathcal{C}_{\eta(\gamma-\tilde{\gamma})}}$ and $O_{\mathcal{D}_{\eta\varepsilon}}$ to solve (RefProb) using fixed precision ξ. If the oracles accept the candidate state, then Algorithm 8 returns a real-valued vector $y^{(k)} \in \mathbb{R}^2$ along with a diagonal matrix $D^{(k)}$ such that the Hamiltonian associated with the Gibbs state that solves the refining problem is

$$H^{(k)} = y_1^{(k)} Q^{(k)} \circ \widetilde{C} + y_2^{(k)} D^{(k)},$$

with $\|y^{(k)}\|_1 \leq 4\log(n)\xi^{-2}$ and $\|D^{(k)}\| \leq 1$ for every $k \geq 0$. This allows us to efficiently describe the solution to each refining problem, and once the algorithm has terminated, it facilitates an efficient way to describe the final solution as well.[2] First, observe that the matrices $Q^{(k)}$ and $D^{(k)}$ can be completely described by their diagonal elements; letting $q^{(k)} \in \mathbb{R}^n$ and $d^{(k)} \in \mathbb{R}^n$ be the vectors that store the diagonal elements of $Q^{(k)}$ and $D^{(k)}$, respectively, we have

$$Q^{(k)} = (ee^\top - I) + \operatorname{diag}\left(q^{(k)}\right),$$
$$D^{(k)} = \operatorname{diag}\left(d^{(k)}\right).$$

Therefore, we store the solution to the refining problem at iteration k as the tuple

$$(\eta^{(k)}, y^{(k)}, q^{(k)}, d^{(k)}, \delta^{(k)}),$$

and the final solution to (OptRelaxed) is defined as

$$\tilde{\rho} = \sum_{k=0}^{K} \frac{1}{\eta^{(k)}(1 + n\delta^{(k)})} \left[Q^{(k)} \circ \frac{\exp\left(-\left[y_1^{(k)} Q^{(k)} \circ \widetilde{C} + y_2^{(k)} \operatorname{diag}\left(d^{(k)}\right)\right]\right)}{\operatorname{tr}\left(\exp\left(-\left[y_1^{(k)} Q^{(k)} \circ \widetilde{C} + y_2^{(k)} \operatorname{diag}\left(d^{(k)}\right)\right]\right)\right)} + \delta^{(k)} I \right].$$

$$(7.24)$$

We point out that this marks a key difference between the output of our algorithm and other quantum SDO solvers based on Gibbs sampling [BKL$^+$19, BKF22, BS17, vAG19, vAGGdW20], which need only return a single state preparation pair. This however does not increase the cost of the method; the iteration bound in Corollary 7.4 ensures that there are only at most $\widetilde{\mathcal{O}}_{\frac{1}{\xi}}(1)$ (i.e., a poly-logarithmic number) of these tuples to be stored over the course of the algorithm. Using the QRAM input model, one can use the stored tuples to construct a block-encoding of the final solution up to error θ using $\widetilde{\mathcal{O}}_{n,\|C\|_F,\frac{1}{\xi},\frac{1}{\theta}}(\sqrt{n})$ queries to the QRAM and $\widetilde{\mathcal{O}}_{n,\|C\|_F,\frac{1}{\xi}}(n)$ classical operations. This construction, and the associated time complexity are analyzed later in Proposition 7.20. We further demonstrate that provided classical access to an s-sparse matrix $A \in \mathbb{R}^{n\times n}$ (with subnormalization factor 1) and access to QRAM, one can estimate $\operatorname{tr}(A\tilde{\rho})$ to additive error θ using $\widetilde{\mathcal{O}}_{n,\|C\|_F,\frac{1}{\xi}}\left(\frac{\sqrt{n}}{\theta}\right)$ queries to the QRAM and $\widetilde{\mathcal{O}}_{n,\|C\|_F,\frac{1}{\xi}}(ns)$ classical

---

[2] Requiring an explicit classical description of the solution would in fact lead to a worse running time overall when compared to the classical implementation we studied in Section 7.5.1.

operations. If A has a subnormalization factor $\alpha_A > 1$, then θ must be scaled down by α_A, increasing the cost.

There is one more important distinction to make between the quantum and classical implementations our of algorithm. When we compute Q in the quantum setting, we do not know the sign of the constraint violations for the diagonal elements exactly. Rather, we only estimate them using our estimate for the diagonal. This stands in contrast with the classical implementation; even though the solution we maintain is approximate, Q is computed *exactly* with respect to the current solution. In the quantum implementation, we only maintain a description of the diagonal elements, and thus do not maintain global information on the current solution. However we can only make a incorrect estimate of the sign if the value of the diagonal element is already very close to the target value, so at most in the refined solution we will "double" this already negligible error.

Additionally, we require Algorithm 8 to return the estimates $\tilde{p}^{(k)} \in \mathbb{R}^n$ (a classical estimate of the diagonal elements of the solution to the refining problem) and $\tilde{c}^{(k)} \in \mathbb{R}$ (a classical estimate of the objective value attained by the solution of the refining problem) that are used to test ξ-closeness for the accepted state. In this fashion, we can (classically) prepare the refining problem data for the next iteration without increasing the cost of the algorithm with respect to n; the objective value can be updated using $\mathcal{O}(1)$ arithmetic operations using $\tilde{c}^{(k)}$, while updating the residuals along the diagonal of ρ requires $\mathcal{O}(n)$ arithmetic operations provided classical access to $\tilde{p}^{(k)}$.

If the current solution is indistinguishable up to precision ζ from the maximally mixed state $n^{-1}I$, and provides an objective value of at least $\gamma - \zeta$, the algorithm terminates and reports the current solution. Otherwise, we construct the refining problem associated with our current solution and proceed to the next iteration.

The next result gives the overall running time required to solve (QUBO-SDO) to additive error $\mathcal{O}(\epsilon)$ using the QRAM input model.

Theorem 7.19. *Let $C \in \mathcal{S}^n$, $\epsilon \in (0,1)$, and set $\zeta = \left(\frac{\epsilon}{n\|C\|_F}\right)^4$. Assume we have classical access to C. Then, in the QRAM input model, Algorithm 10 solves (QUBO-SDO) up to additive error $\mathcal{O}(\epsilon)$ using*

$$\mathcal{O}\left(n^{1.5} \cdot \operatorname{polylog}\left(n, \|C\|_F, \frac{1}{\epsilon}\right)\right)$$

accesses to the QRAM and $\mathcal{O}(ns)$ classical arithmetic operations.

The output of the algorithm is a collection of tuples $\{(\eta^{(k)}, y^{(k)}, q^{(k)}, d^{(k)}, \delta^{(k)})\}_{k=0}^{K}$ such that

$$\tilde{\rho} = \sum_{k=0}^{K} \frac{1}{\eta^{(k)}\left(1 + n\delta^{(k)}\right)}\left[Q^{(k)} \circ \widetilde{C}\frac{\exp\left(-\left[y_1^{(k)} Q^{(k)} \circ \widetilde{C} + y_2^{(k)}\operatorname{diag}\left(d^{(k)}\right)\right]\right)}{\operatorname{tr}\left(\exp\left(-\left[y_1^{(k)} Q^{(k)} \circ \widetilde{C} + y_2^{(k)}\operatorname{diag}\left(d^{(k)}\right)\right]\right)\right)} + \delta^{(k)}I\right] \succeq 0,$$

is a ζ-precise solution to (OptRelaxed). The entries of $\tilde{\rho}$ can be modified to construct a matrix

---

Algorithm 10 Iterative Refinement for SDO Approximations of QUBOs using a quantum computer

---

Input: Error tolerances $\epsilon \in (0,1)$ and $\zeta = \left(\frac{\epsilon}{n\|C\|_F}\right)^4$, upper bound on objective value $\gamma \in [-1,1]$

Output: Tuples $\left\{\left(\eta^{(k)}, y^{(k)}, q^{(k)}, d^{(k)}, \delta^{(k)}\right)\right\}_{k=0}^{K}$ that define a ζ-precise solution $\tilde{\rho}$ to (OptRelaxed) using Equation (7.24)

Initialize: $\tilde{p}, \hat{p} \leftarrow \mathbf{0}^n$, $Q \leftarrow ee^\top$, $\varepsilon_i = \frac{1}{n}$ for $i \in [n]$, $\tilde{\gamma}, \hat{\gamma} \leftarrow 0$, $\eta^{(0)} \leftarrow 1$, $\delta^{(0)} \leftarrow 0$, $k \leftarrow 1$

$(y^{(0)}, D^{(0)}, \tilde{p}^{(0)}, \tilde{c}^{(0)}) \leftarrow$ solve (RefProb) to precision $\frac{\xi^2}{16}$ using Algorithm 8 with oracles $O_{\mathcal{C}_{\eta(\gamma-\tilde{\gamma})}}$ and $O_{\mathcal{D}_{\eta\varepsilon}}$

$\tilde{\gamma}^{(0)} \leftarrow c^{(0)}$

$\varepsilon_i^{(0)} \leftarrow \tilde{p}_i^{(0)} - \frac{1}{n}$ for $i \in [n]$

$Q_{ii} \leftarrow \text{sign}(-\varepsilon_i^{(0)})$ for $i \in [n]$

$\eta^{(1)} \leftarrow \frac{1}{\max\left\{\gamma - \tilde{\gamma}^{(0)}, \|\varepsilon^{(0)}\|_1\right\}}$

$\delta^{(1)} \leftarrow \frac{2}{n}\left(\|\varepsilon^{(0)}\|_1 + \frac{(\xi/4)^2}{\eta^{(1)}}\right)$

while $\max\left\{\gamma - \tilde{\gamma}, \|\varepsilon\|_1\right\} > \zeta$ **do**

1. Store refining problem data $\left(Q \circ \widetilde{C}, \eta^{(k)}\varepsilon^{(k-1)}, \eta^{(k)}\tilde{\gamma}^{(k-1)}\right)$

2. $(y^{(k)}, D^{(k)}, p^{(k)}, \tilde{c}^{(k)}) \leftarrow$ Solve (RefProb) to precision $\frac{\xi^2}{16}$ using Algorithm 8

3. Update estimate of diagonal entries
$$\hat{p}_i^{(k)} \leftarrow \hat{p}_i^{(k-1)} + \frac{Q_{ii}}{\eta^{(k)}}p_i^{(k)} \quad \text{for } i \in [n]$$

4. Apply spectrum shift to estimate of diagonal entries and update objective value
$$\hat{p}_i^{(k)} \leftarrow \frac{1}{1+n\delta^{(k)}}\left(\hat{p}_i^{(k)} + \delta^{(k)}\right) \quad \text{for } i \in [n], \quad \tilde{\gamma}^{(k)} \leftarrow \tilde{\gamma}^{(k-1)} + \frac{1}{\eta^{(k)}}\left[\frac{1}{1+n\delta^{(k)}}\left(\tilde{c}^{(k)} + \delta^{(k)}\operatorname{tr}(\widetilde{C})\right)\right]$$

5. Store diagonal elements of Q and $D^{(k)}$ as the vectors $\left(q^{(k)}, d^{(k)}\right) \in \mathbb{R}^n \times \mathbb{R}^n$

6. Store description of solution to the refining problem $(\eta^{(k)}, y^{(k)}, q^{(k)}, d^{(k)}, \delta^{(k)})$

7. Compute element-wise deviations from the maximally mixed state:
$$\varepsilon_i^{(k)} \leftarrow \hat{p}_i^{(k)} - \frac{1}{n} \quad \text{for } i \in [n]$$

8. Classically update refining problem parameters:
$$Q_{ii} \leftarrow \text{sign}\left(-\varepsilon_i^{(k)}\right) \text{ for } i \in [n], \quad \eta^{(k+1)} \leftarrow \frac{1}{\max\left\{\gamma - \tilde{\gamma}^{(k)}, \|\varepsilon^{(k)}\|_1\right\}}$$

9. Classically update spectrum shift parameter:
$$\delta^{(k+1)} \leftarrow \frac{2}{n}\left(\left\|\varepsilon^{(k)}\right\|_1 + \frac{(\xi/4)^2}{\eta^{(k+1)}}\right)$$

10. $k \leftarrow k + 1$

end

ρ^* at trace distance $\mathcal{O}\left(\frac{\epsilon}{n\|C\|_F}\right)$ of $\tilde{\rho}$ in time $\mathcal{O}(n^2)$, such that $n\rho^*$ is a feasible point of the SDO problem (QUBO-SDO).

Proof. Given that C is an s-sparse matrix, we can classically load C in $\mathcal{O}(ns)$ time. Similarly, for normalization purposes we classically compute $\|C\|_F$, which requires $\mathcal{O}(ns)$ arithmetic operations. In each iteration we use Algorithm 8 to solve (RefProb), and use classical estimates of the diagonal elements of the refining solution, and a classical estimate of the objective value attained by the refining solution to update the solution and data for the refining problem we need to solve in the next iteration.

Let $\mathcal{T}_{HU}^{\text{quantum}}$ denote the cost of using Algorithm 8 as an approximate SDO subroutine at each iterate of Algorithm 10. By Proposition 7.14, Algorithm 8 solves (RefProb) to additive error ξ using at most

$$\tilde{\mathcal{O}}_{\frac{n}{\xi}}\left(n^{1.5}\xi^{-5}\right)$$

accesses to the QRAM. In the context of Algorithm 10, ξ is a fixed constant (and hence, so is ξ^2), so each ξ^2-precse oracle call to Algorithm 8 has an associated cost of

$$\mathcal{T}_{HU}^{\text{quantum}} = \tilde{\mathcal{O}}_n\left(n^{1.5}\right)$$

accesses to the QRAM.

Classically updating the objective value requires $\mathcal{O}(1)$ arithmetic operations while updating the vector p which stores a classical description of the diagonal elements of our solution requires $\mathcal{O}(n)$ classical arithmetic operations. Similarly, accounting for the spectrum shift requires $\mathcal{O}(n)$ arithmetic operations; again this step is limited to operations on n-dimensional vectors and computing the trace of $\tilde{C}$ (which can be stored once at the start of the algorithm). Likewise, ε and Q can each be updated using $\mathcal{O}(n)$ classical arithmetic operations, as we only need to store the diagonal elements of Q. This also implies that we can update $Q \circ \tilde{C}$ using $\tilde{\mathcal{O}}_n(n)$ operations, for only the diagonal elements need to be updated. When compared to loading and normalizing the coefficient matrix C, or our use of Algorithm 8 as a subroutine for solving (RefProb), these intermediate computation steps are negligible and do not factor into the overall running time using $\mathcal{O}$ notation.

By Corollary 7.4, Algorithm 10 terminates in at most $\tilde{\mathcal{O}}_{\frac{1}{\zeta}}(1)$ iterations. Therefore, the worst case complexity of Algorithm 10 can be bounded by

$$\mathcal{O}\left(n^{1.5} \cdot \text{polylog}\left(n, \|C\|_F, \frac{1}{\epsilon}\right)\right)$$

accesses to the QRAM, and $\mathcal{O}(ns)$ classical arithmetic operations. Just as in the proof of Theorem 7.18, applying Proposition 7.17 with our choice of $\zeta = \left(\frac{\epsilon}{n\|C\|_F}\right)^4$ implies that the above running time is sufficient to obtain a solution that can be used to solve (QUBO-SDO) up to additive error $\mathcal{O}(\epsilon)$, and the proof is complete. $\qquad\square$

Using the sparse-access input model, one can show that the resulting scheme exhibits an oracle complexity of

$$\mathcal{O}\left(n^{1.5}s^{0.5+o(1)} \cdot \mathrm{polylog}\left(n, \|C\|_F, \frac{1}{\epsilon}\right)\right),$$

and requires $\mathcal{O}\left(n^{2.5}s^{0.5+o(1)} \cdot \mathrm{polylog}\left(n, \|C\|_F, \frac{1}{\epsilon}\right)\right)$ additional gates. To summarize, in the absence of QRAM, the number of oracle accesses is a factor $\sqrt{s}$ larger due to the Hamiltonian simulation, and the gate complexity increases by a factor n due to the cost of constructing O_D without QRAM.

We conclude this section by establishing the costs of preparing a block-encoding of the final solution, and estimating trace inner products of the form $\mathrm{tr}(A\tilde{\rho})$ for a given matrix A.

Proposition 7.20. *Suppose that Algorithm 10 is run with $\zeta = \left(\frac{\epsilon}{n\|C\|_F}\right)^4$ for some $\epsilon \in (0,1)$, and terminates after K iterations, classically outputting the tuples $\left\{\left(\eta^{(k)}, y^{(k)}, q^{(k)}, d^{(k)}, \delta^{(k)}\right)\right\}_{k=0}^{K}$. Then, letting $\widetilde{C} = C\|C\|_F^{-1}$ be stored in QRAM, and denoting the refining problem at iteration k by $\rho^{(k)}$, one can use $\{(\eta^{(k)}, y^{(k)}, q^{(k)}, d^{(k)}, \delta^{(k)})\}_{k=0}^{K}$ to implement an $(n, \mathcal{O}(\log(n)), \theta)$-block-encoding of*

$$\tilde{\rho} = \sum_{k=0}^{K} \frac{1}{\eta^{(k)}\left(1 + n\delta^{(k)}\right)}\left[Q^{(k)} \circ \frac{\exp\left(-\left[y_1^{(k)}Q^{(k)} \circ \widetilde{C} + y_2^{(k)}\,\mathrm{diag}\left(d^{(k)}\right)\right]\right)}{\mathrm{tr}\left(\exp\left(-\left[y_1^{(k)}Q^{(k)} \circ \widetilde{C} + y_2^{(k)}\,\mathrm{diag}\left(d^{(k)}\right)\right]\right)\right)} + \delta^{(k)}I \right],$$

with at most $\widetilde{\mathcal{O}}_{n,\|C\|_F,\frac{1}{\epsilon},\frac{1}{\theta}}\left(\sqrt{n}\right)$ queries to the QRAM and $\widetilde{\mathcal{O}}_{n,\|C\|_F,\frac{1}{\epsilon},\frac{1}{\theta}}(n)$ classical operations.

Proof. First, note that

$$\frac{\exp\left(-\left[y_1^{(k)}Q^{(k)} \circ \widetilde{C} + y_2^{(k)}\,\mathrm{diag}\left(d^{(k)}\right)\right]\right)}{\mathrm{tr}\left(\exp\left(-\left[y_1^{(k)}Q^{(k)} \circ \widetilde{C} + y_2^{(k)}\,\mathrm{diag}\left(d^{(k)}\right)\right]\right)\right)} = n^{-1}I$$

whenever $y = (0,0)^\top$. Thus, by choosing

$$\Delta := \sum_{k=1}^{K} \frac{\delta^{(k)}}{\eta^{(k)}\left(1 + n\delta^{(k)}\right)},$$

and setting $y^{(K+1)} = (0,0)^\top$, $\delta^{(K+1)} = \frac{1}{n}$, $\eta^{(K+1)} = \frac{1}{n\Delta}$, and $Q^{(K+1)} = ee^\top$ we can simplify the expression of the final solution to

$$\tilde{\rho} = \sum_{k=0}^{K+1} \frac{1}{\eta^{(k)}\left(1 + n\delta^{(k)}\right)}Q^{(k)} \circ \frac{\exp\left(-\left[y_1^{(k)}Q^{(k)} \circ \widetilde{C} + y_2^{(k)}\,\mathrm{diag}\left(d^{(k)}\right)\right]\right)}{\mathrm{tr}\left(\exp\left(-\left[y_1^{(k)}Q^{(k)} \circ \widetilde{C} + y_2^{(k)}\,\mathrm{diag}\left(d^{(k)}\right)\right]\right)\right)}.$$

To ensure that the stated complexity holds, for each $k \in [K+1]$, we block-encode

$$A^{(k)} = Q^{(k)} \circ \widetilde{C} + D^{(k)}.$$

First, note that with classical access to C and $q^{(k)}$, one can store $Q^{(k)} \circ \widetilde{C}$ in the QRAM by properly updating $\widetilde{C}$ in the QRAM. This step requires $\mathcal{O}(n)$ classical operations, as the only non-trivial computation that is performed is limited to the diagonal elements of the involved matrices. Then, with $Q^{(k)} \circ \widetilde{C}$ stored in QRAM, noting that $\left\| Q^{(k)} \circ \widetilde{C} \right\|_F \leq 1$ holds for every $k \in [K+1]$, we apply Lemma 3.3 to construct a $(1, \log(n) + 2, \theta_1)$-block-encoding of $Q^{(k)} \circ \widetilde{C}$ in time $\mathcal{O}\left(\mathrm{polylog}\left(\frac{n}{\theta_1} \right) \right)$. Similarly, as we saw in the proof of Proposition 7.9, classical access to $d^{(k)}$ and access to QRAM implies one can implement a $(1, \log(n) + 3, \theta_1)$-block-encoding of $D^{(k)}$ can be constructed in time $\widetilde{\mathcal{O}}_{\frac{n}{\theta_1}}(1)$.

Again following the proof of Proposition 7.9, applying Corollary 7.2 with $y^{(k)}$ satisfying $\|y^{(k)}\|_1 = \widetilde{\mathcal{O}}_n(\xi^{-1})$ implies that we can construct a unitary which prepares a copy of the Gibbs state $\rho^{(k)}$ encoding the solution to the refining problem at iteration k with at most

$$\widetilde{\mathcal{O}}_{\frac{n}{\xi}}\left(\sqrt{n}\alpha\xi^{-1} \right) = \widetilde{\mathcal{O}}_n\left(\sqrt{n} \right),$$

accesses to the QRAM, as $\alpha = 1$ and ξ is a fixed constant. Therefore, by Lemma 3.18, preparing a $(1, \log(n) + a, \theta_1)$ block-encoding of a purification of $\rho^{(k)}$ thus requires $\widetilde{\mathcal{O}}_{\frac{n}{\theta_1}}(\sqrt{n})$ queries to the QRAM.

Next, provided classical access to the vector $q^{(k)}$ that store the diagonal elements of $Q^{(k)}$, access to QRAM implies that we can efficiently implement an oracle $O_{Q^{(k)}}$ that returns the entries of $Q^{(k)}$ in a binary description:

$$O_{Q^{(k)}} : |i\rangle|j\rangle|0\rangle^{\otimes p} \mapsto |i\rangle|j\rangle|q_{ij}^{(k)}\rangle, \quad \forall i,j \in [2^{\log n}] - 1,$$

where $q_{ij}^{(k)}$ is a p-bit binary description of the ij-matrix element of $Q^{(k)}$ for $k = 0, \ldots, K+1$. By construction each matrix $Q^{(k)}$ may be fully dense, and hence an application of Lemma 3.2 with $s_r = s_c = n$ asserts that in the presence of QRAM, one can construct a $(n, \log(n) + 3, \theta_2)$-block-encoding of $Q^{(k)}$ in time $\widetilde{\mathcal{O}}_{\frac{n}{\theta_2}}(1)$.

From here, we can utilize Proposition 3.11 with $\theta_1 = \theta_2 = \frac{\tilde{\theta}}{10}$ to construct an $(n, a + 4\log(n^2) + 12, \tilde{\theta})$-block-encoding of $Q^{(k)} \circ \rho^{(k)}$ in time $\widetilde{\mathcal{O}}_{\frac{n}{\tilde{\theta}}}(1)$. Repeating the above steps for $k = 0, \ldots, K+1$, it follows that we can block-encode each of the terms $Q^{(k)} \circ \rho^{(k)}$ using at most

$$\widetilde{\mathcal{O}}_{n, \frac{1}{\tilde{\theta}}}\left(K\sqrt{n} \right) = \widetilde{\mathcal{O}}_{n, \|C\|_F, \frac{1}{\epsilon}, \frac{1}{\tilde{\theta}}}\left(\sqrt{n} \right)$$

queries to the QRAM and $\widetilde{\mathcal{O}}_{n, \|C\|_F, \frac{1}{\epsilon}, \frac{1}{\tilde{\theta}}}(n)$ classical operations, as $K = \mathcal{O}\left(\mathrm{polylog}\left(n, \|C\|_F, \frac{1}{\epsilon} \right) \right)$ by Corollary 7.4.

Finally, what remains is to take the linear combination of these terms. To do so, we choose our weights to be $w_k = \frac{1}{2(1+n\delta^{(k)})\eta^{(k)}}$, which indeed satisfies $\|w\|_1 \leq 1$. Then, we can construct a $(K+2, \log(K+2), 0)$-state-preparation pair P_L, P_R for w, which can be constructed by taking a $\log(K+2)$-fold tensor product of the Hadamard gate, i.e.,

$$P_L = P_R = \frac{1}{\sqrt{2}} \begin{pmatrix} 1 & 1 \\ 1 & -1 \end{pmatrix}^{\otimes \log(K+2)}.$$

We are now in a position to apply Proposition 3.4, and choosing $\tilde{\theta} = \frac{\theta}{n}$, we can obtain W upon adding a control qubit to the circuits used to construct the block-encoding of each $Q^{(k)} \circ \rho^{(k)}$. As a result, we obtain an $(n, \mathcal{O}(\log(n)), \theta)$-block-encoding of $\tilde{\rho}$ with a single use of W, P_R and $P_L^\dagger$. Summing the cost of each step in the construction we arrive at total cost of

$$\tilde{\mathcal{O}}_{n, \|C\|_F, \frac{1}{\epsilon}, \frac{1}{\theta}} \left(\sqrt{n} \right)$$

queries to the QRAM and $\tilde{\mathcal{O}}_{n, \|C\|_F, \frac{1}{\epsilon}, \frac{1}{\theta}} (n)$ classical operations, and proof is complete. $\square$

Proposition 7.21. *Suppose that Algorithm 10 is run with $\zeta = \left(\frac{\epsilon}{n\|C\|_F} \right)^4$ for some $\epsilon \in (0, 1)$, and terminates after K iterations, classically outputting the tuples $\left\{ \left(\eta^{(k)}, y^{(k)}, q^{(k)}, d^{(k)}, \delta^{(k)} \right) \right\}_{k=0}^{K}$. Let $A \in \mathbb{R}^{n \times n}$ be a matrix with $\|A\|_F \leq 1$ and assume classical access to A and $C/\|C\|_F$. Then, with access to QRAM, one can compute a θ-precise estimate of $\mathrm{tr}(A\tilde{\rho})$ using at most $\tilde{\mathcal{O}}_{n, \|C\|_F, \frac{1}{\epsilon}} \left(\frac{\sqrt{n}}{\theta} \right)$ queries to the QRAM and $\tilde{\mathcal{O}}_{n, \|C\|_F, \frac{1}{\epsilon}} (n)$ classical operations.*

Proof. See the proof of Theorem 7.25 in Section 7.6. $\square$

A QRAM-free version of Proposition 7.21 is also analyzed in Section 7.6, and the cost is summarized in Corollary 7.5. Without access to QRAM, the cost increases with respect to n because computing the Hadamard product of block-encodings introduces n as a subnormalization factor. This is compounded in the running time, upon noting that we then have to scale down the error for the amplitude estimation steps by n, and constructing sparse-access oracles for the intermediate block-encodings of Q and D that arise in the trace estimation procedure requires $\tilde{\mathcal{O}}_n(n)$ gates.

7.5.3 Comparison to existing SDO algorithms

Table 7.1 presents a comparison of the running time results for the algorithms we have proposed with the running times of the best performing methods from both the classical and quantum literature when applied to solving (QUBO-SDO).

Note that when directly solving (QUBO-SDO), $m = n$, and any feasible solution X to (QUBO-SDO) satisfies $\mathrm{tr}(X) = n$, implying $R = n$ for the algorithms based on the (Q)MMWU framework. We also point out that the running times in Table 7.1 take into account the role of sparsity in context of the algorithms, which is measured as the maximum number of nonzero entries per row of the constraint matrices $A_1, \ldots, A_n$. When using either an IPM or CPM to

References	Method	Runtime	Error Scaling
[JKL$^+$20]	IPM	$\mathcal{O}_{n,\frac{1}{\epsilon}}\left(n^{\omega+0.5}\right)$	ϵ
[ANTZ21]	QIPM	$\tilde{\mathcal{O}}_{n,\kappa,\frac{1}{\epsilon}}\left(\sqrt{n}(n^3\kappa\epsilon^{-1}+n^4)\right)$	ϵ
[LP20]	MMWU	$\tilde{\mathcal{O}}_{n,\frac{1}{\epsilon}}\left(ns\epsilon^{-3.5}\right)$	$\|C\|_{\ell_1}\epsilon$
[vAG19]	QMMWU	$\tilde{\mathcal{O}}_{n,\frac{1}{\epsilon}}\left(n^{5.5}s\epsilon^{-4}\right)$	$n\|C\|\epsilon$
[BKF22] (Classical)	HU	$\tilde{\mathcal{O}}_{n,\|C\|}\left(\min\{n^2s,n^\omega\}\epsilon^{-12}\right)$	$n\|C\|\epsilon$
[BKF22] (Quantum)	HU	$\tilde{\mathcal{O}}_{n,\|C\|,\frac{1}{\epsilon}}\left(n^{2.5}s^{0.5+o(1)}\epsilon^{-28+o(1)}\exp\left(1.6\sqrt{\log(\epsilon^{-1})}\right)\right)$	$n\|C\|\epsilon$
[BKF22] (Quantum)	HU-QRAM	$\tilde{\mathcal{O}}_{n,\|C\|,\frac{1}{\epsilon}}\left(n^{1.5}s^{0.5+o(1)}\epsilon^{-28+o(1)}\exp\left(1.6\sqrt{\log(\epsilon^{-1})}\right)\right)$	$n\|C\|\epsilon$
This work (Classical)	IR-HU	$\tilde{\mathcal{O}}_{n,\|C\|_F,\frac{1}{\epsilon}}\left(\min\{n^2s,n^\omega\}\right)$	ϵ
This work (Quantum)	IR-HU	$\tilde{\mathcal{O}}_{n,\|C\|_F,\frac{1}{\epsilon}}\left(n^{2.5}s^{0.5+o(1)}\right)$	ϵ
This work (Quantum)	IR-HU-QRAM	$\tilde{\mathcal{O}}_{n,\|C\|_F,\frac{1}{\epsilon}}(n^{1.5})+ns$	ϵ

Table 7.1: Total running times for classical and quantum algorithms to solve (QUBO-SDO).

solve (QUBO-SDO), the n constraint matrices are $A_i = e_ie_i^\top$ (with row sparsity one) enforcing $X_{ii} = 1$ for each diagonal element. On the other hand, algorithms based on the (Q)MMWU or HU frameworks solve (QUBO-SDO) by reducing the problem to a feasibility problem; C enters into the resulting formulation as another constraint matrix, and as a result, the relevant sparsity parameter is the maximum number of non-zeroes per row of C, which we denote by s in Table 7.1.

There are additional considerations that need to be taken into account when making comparisons across methodologies listed in Table 7.1. Broadly speaking, both (Q)MMWUs and HU require normalizing the problem by an upper bound on the trace of a primal solution, and in the case of (QUBO-SDO), we have the natural bound $\text{tr}(X) = n$. Moreover, (Q)MMWUs and HU additionally normalize the cost matrix so that it exhibits unit norm with respect to some norm. While these modifications amount to scaling the optimal objective value of (QUBO-SDO) by a fixed quantity, without employing any safeguards such as IR, these modifications impact the scaling of the error as reflected in the fourth column of Table 7.1. On the contrary, (Q)IPMs do not require the SDO problem to be normalized in any way. Finally there is a distinction with regard to output; (Q)IPMs explicitly report a classical description of the solution X, whereas only the classical HU algorithm of [BKF22] and our own classical IR-HU method do so; the primal QMMWU of [vAG19] reports a state-preparation pair y, and the MMWU algorithm found in [LP20] reports a "gradient" $G \in \mathcal{S}^n$ such that $X = W\exp(G)W$ for a diagonal matrix W. As we noted earlier, (Q)IPMs and (Q)MMWUs also utilize different definitions of optimality.

It can be easily seen that both the classical and quantum implementations of our proposed methodology outperform all existing algorithms that exhibit poly-logarithmic dependence on the precision ϵ. Our classical algorithm is only outperformed with respect to dimension by our own quantum algorithms, and the algorithm from [LP20], which has an exponentially worse dependence on the inverse prevision. Moreover, to achieve the same error scaling as our algorithms, the algorithm from [LP20] would require time $\tilde{\mathcal{O}}_{n,\frac{1}{\epsilon}}\left(\|C\|_{\ell_1}^{3.5}ns\epsilon^{-3.5}\right)$. Up to poly-logarithmic factors, our quantum algorithms outperform each of the classical and quantum solvers in every parameter, suggesting the first evidence of quantum advantage for solving a special class of SDO problems. Moreover, our implementation with access to QRAM dominates all other algorithms. We therefore conclude that our proposed algorithms are respectively, the fastest both in the classical and quantum regimes.

7.6 Additional results

7.6.1 Running time of Algorithm 10 without QRAM

The following result from [BKF22] gives the sample complexity of implementing the oracles in the sparse-access model.

Lemma 7.22 (see, proof of Lemma 3.3 in [BKF22]). *We can implement the oracle $O_{\mathcal{C}_\gamma}$ on a quantum computer given access to $\mathcal{O}(\epsilon^{-2})$ copies of a state that is an $\frac{\epsilon}{8}$-approximation of the input state ρ in trace distance. The oracle $O_{\mathcal{D}_n}$ can be implemented using $\mathcal{O}(n\epsilon^{-2})$ $\frac{\epsilon}{8}$-approximate copies of the input, and the classical post-processing time needed to implement the oracle is $\mathcal{O}(n\epsilon^{-2})$.*

Next, we bound the overall complexity of Algorithm 8 without access to QRAM.

Proposition 7.23. *Suppose that $C \in \mathcal{S}^n$ has row sparsity s and $\xi \in (0,1)$. Then, in the sparse-access input model, the complexity of solving (OptRelaxed) up to additive error ξ using Algorithm 8 on a quantum computer requires*

$$\tilde{\mathcal{O}}_n\left(n^{1.5}\sqrt{s}^{\,1+o(1)}\xi^{-7+o(1)}\exp\left(1.6\sqrt{\log(\xi^{-1})}\right)\right)$$

queries to the input oracle O_C and $\tilde{\mathcal{O}}_n\left(n^{2.5}\sqrt{s}^{\,1+o(1)}\xi^{-7+o(1)}\exp\left(1.6\sqrt{\log(\xi^{-1})}\right)\right)$ additional gates.

Proof. Our proof can be viewed as the QRAM-free analogue of the discussion found in [BKF22, Section 3.4], and we repeat it here for completeness. In order to derive an appropriate bound on the per-iteration cost, we need to evaluate the cost of constructing our separation oracles. By Lemma 7.22, we can conclude that the time to construct the oracle $O_{\mathcal{D}_n}$ for the diagonal elements dominates that of constructing the oracle $O_{\mathcal{C}_\gamma}$ to test the objective value.

We now turn our attention to the cost of simulating our Hamiltonian H. From the results in [PW09, Appendix] it follows that we can produce a state that is $\frac{\epsilon}{8}$ close to ρ using $\tilde{\mathcal{O}}(\sqrt{n}\xi^{-3})$ invocations of a controlled U which satisfies

$$\left\|U - e^{it_0 H}\right\| \leq \mathcal{O}\left(\xi^3\right),$$

with $t_0 = \frac{\pi}{4\|H\|}$. Further, the authors in [BKF22] note that each of the Hamiltonians we seek to simulate are of the form $H = y_1 C\|C\|_F^{-1} + y_2 D$ where $y_1, y_2 = \mathcal{O}(\log(n)\xi^{-1})$ and D is a diagonal matrix which satisfies $\|D\| \leq 1$. Invoking [CW12, Theorem 1], we can simulate H for time t up to error ξ^3 using

$$\tilde{\mathcal{O}}\left(t(a+b)\exp\left(1.6\sqrt{\log\left(\log(n)t\xi^{-3}\right)}\right)\right)$$

separate simulations of $y_1C\|C\|_F$ and y_2D.

As noted in [BKF22], access to the oracles O_{sparse} and O_C we described in Section 2.5.6 allows us to simulate $\exp(it\tilde{C})$ in time $\mathcal{O}\left((t\sqrt{s})^{1+o(1)}\xi^{o(1)}\right)$ if we utilize the algorithm in [Low19]. Similarly, we follow [BKF22] in constructing an oracle O_D acting on $\mathbb{C}\otimes(\mathbb{C}^2)^{\otimes a}$, where a is a sufficiently large constant such that we can represent the diagonal elements of D as

$$O_D|i,z\rangle \mapsto |i, z\oplus D_{ii}\rangle$$

to the desired level of precision in binary. Accordingly, we can simulate e^{iDt} for $t = \tilde{\mathcal{O}}(\xi^{-1})$ using $\tilde{\mathcal{O}}_n(1)$ queries to O_D and $\tilde{\mathcal{O}}_n(1)$ elementary operations [BACS07], and we can implement O_D using $\tilde{\mathcal{O}}_n(n)$ gates.

To summarize, the Gibbs sampler from [PW09] requires $\tilde{\mathcal{O}}(\sqrt{n}\xi^{-3})$ Hamiltonian simulation steps, each of which requires time

$$\tilde{\mathcal{O}}\left(\sqrt{s}^{1+o(1)}\xi^{o(1)}\exp\left(1.6\sqrt{\log(\xi^{-1})}\right)\right).$$

Hence, each iteration of Algorithm 8 requires a total of

$$\tilde{\mathcal{O}}_n\left(n^{1.5}\sqrt{s}^{1+o(1)}\xi^{-5+o(1)}\exp\left(1.6\sqrt{\log(\xi^{-1})}\right)\right)$$

sparse-access oracle queries. Combining the above per-iteration cost with the iteration bound $\mathcal{O}(\log(n)\xi^{-2})$ provided in Theorem 7.3, it follows that Algorithm 8 solves (OptRelaxed) up to additive error ξ with at most

$$\tilde{\mathcal{O}}_n\left(n^{1.5}\sqrt{s}^{1+o(1)}\xi^{-7+o(1)}\exp\left(1.6\sqrt{\log(\xi^{-1})}\right)\right)$$

queries to the input oracle O_C and $\tilde{\mathcal{O}}_n\left(n^{2.5}\sqrt{s}^{1+o(1)}\xi^{-7+o(1)}\exp\left(1.6\sqrt{\log(\xi^{-1})}\right)\right)$ additional gates. $\qquad\square$

Theorem 7.24 formalizes the complexity of of Algorithm 10 in the quantum setting without access to QRAM. In our analysis, we employ the same Hamiltonian simulation subroutines and Gibbs sampler used in [BKF22] to construct our separation oracles.

Theorem 7.24. *Let $C \in \mathcal{S}^n$ with row sparsity s and $\epsilon \in (0,1)$. Then, setting $\zeta = \left(\frac{\epsilon}{n\|C\|_F}\right)^4$ and fixing $\xi = 10^{-2}$, a quantum implementation of Algorithm 10 using the sparse-access input model solves (QUBO-SDO) up to additive error $\mathcal{O}(\epsilon)$ using*

$$\mathcal{O}\left(n^{1.5}s^{0.5+o(1)}\cdot\text{polylog}\left(n,\|C\|_F,\frac{1}{\epsilon}\right)\right)$$

queries to the input oracle O_C and $\mathcal{O}\left(n^{2.5}s^{0.5+o(1)} \cdot \mathrm{polylog}\left(n, \|C\|_F, \frac{1}{\epsilon}\right)\right)$ additional gates.

The output of the algorithm is a collection of tuples $\{(\eta^{(k)}, y^{(k)}, q^{(k)}, d^{(k)}, \delta^{(k)})\}_{k=0}^{K}$ such that

$$\tilde{\rho} = \sum_{k=0}^{K} \frac{1}{\eta^{(k)}(1+n\delta^{(k)})} \left[Q^{(k)} \circ \frac{\exp\left(-\left[y_1^{(k)} Q^{(k)} \circ \tilde{C} + y_2^{(k)} \mathrm{diag}\left(d^{(k)}\right)\right]\right)}{\mathrm{tr}\left(\exp\left(-\left[y_1^{(k)} Q^{(k)} \circ \tilde{C} + y_2^{(k)} \mathrm{diag}\left(d^{(k)}\right)\right]\right)\right)} + \delta^{(k)} I \right] \succeq 0,$$

is a ζ-precise solution to (OptRelaxed). The entries of $\tilde{\rho}$ can be modified to construct a matrix ρ^* at trace distance $\mathcal{O}\left(\frac{\epsilon}{n\|C\|_F}\right)$ of $\tilde{\rho}$ in time $\mathcal{O}(n^2)$, such that $n\rho^*$ is a feasible point of the SDO problem (QUBO-SDO).

Proof. Given that C is an s-sparse matrix, we can load C in $\mathcal{O}(ns)$ time. Similarly, for normalization purposes we classically compute $\|C\|_F$, which requires $\mathcal{O}(ns)$ arithmetic operations. In each iteration we use Algorithm 8 to solve (RefProb), and use classical estimates of the diagonal elements of the refining solution, and a classical estimate of the objective value attained by the refining solution to update the solution and data for the refining problem we need to solve in the next iteration.

Letting $\mathcal{T}_{HU}^{\mathrm{sparse}}$ be the cost of using Algorithm 8 as an approximate SDO subroutine, we saw in Proposition 7.23, Algorithm 8 solves (RefProb) to additive error ξ using

$$\mathcal{T}_{HU}^{\mathrm{sparse}} = \tilde{\mathcal{O}}_n \left(n^{1.5}\sqrt{s}^{1+o(1)} \xi^{-7+o(1)} \exp\left(1.6\sqrt{\log(\xi^{-1})}\right) \right)$$

queries to the oracle describing the problem data and $\tilde{\mathcal{O}}_n \left(n^{2.5}\sqrt{s}^{1+o(1)} \xi^{-7+o(1)} \exp\left(1.6\sqrt{\log(\xi^{-1})}\right) \right)$ additional gates. In the context of Algorithm 10, ξ is a fixed constant, so the cost of our oracle call to Algorithm 8 simplifies to

$$\mathcal{T}_{HU}^{\mathrm{sparse}} = \tilde{\mathcal{O}}_n \left(n^{1.5}\sqrt{s}^{1+o(1)} \right)$$

queries to the oracle describing the problem data and $\tilde{\mathcal{O}}_n \left(n^{2.5}\sqrt{s}^{1+o(1)} \right)$ additional gates.

Classically updating the objective value requires $\mathcal{O}(1)$ arithmetic operations while updating the vector p which stores a classical description of the diagonal elements of our solution as

$$p_i \leftarrow p_i + \frac{Q_{ii}}{\eta^{(k)}} \tilde{p}_i^{(k)}$$

requires $\mathcal{O}(n)$ arithmetic operations. Again, ϵ and Q can each be updated using $\mathcal{O}(n)$ arithmetic operations, as we only need to store the diagonal elements of Q. This also implies that we can also calculate $Q \circ \tilde{C}$ in time $\mathcal{O}(n)$, for only the element-wise products along the diagonal are non-trivial. When compared to loading and normalizing the data or our use of Algorithm 8 as a subroutine for solving (RefProb), these intermediate computation steps are negligible and do not factor into the overall running time using $\mathcal{O}$ notation.

Factoring in the $\mathcal{O}\left(\mathrm{polylog}\left(\frac{1}{\zeta}\right)\right) = \mathcal{O}\left(\mathrm{polylog}\left(n, \|C\|_F, \frac{1}{\epsilon}\right)\right)$ from Corollary 7.4, it follows that a quantum implementation of Algorithm 10 requires at most

$$\mathcal{O}\left(n^{1.5}s^{0.5+o(1)} \cdot \mathrm{polylog}\left(n, \|C\|_F, \frac{1}{\epsilon}\right)\right)$$

queries to the input oracle O_C and $\mathcal{O}\left(n^{2.5}s^{0.5+o(1)} \cdot \mathrm{polylog}\left(n, \|C\|_F, \frac{1}{\epsilon}\right)\right)$ additional gates. Just as in the proof of Theorem 7.18, applying Proposition 7.17 with our choice of $\zeta = \left(\frac{\epsilon}{n\|C\|_F}\right)^4$ implies that the above running time is sufficient to obtain a solution that can be used to solve (QUBO-SDO) up to additive error $\mathcal{O}(\epsilon)$, and the proof is complete. $\qquad\square$

7.6.2 Estimating trace inner products with the final solution

Given that we do not explicitly report a classical description of the final solution $\tilde{\rho}$ defined in equation (7.24), it may be of interest to understand how, for a user specified matrix A, one can compute the trace inner product $\mathrm{tr}(A\tilde{\rho})$. We outline a procedure for doing so using the state preparation pair description of solution $\{(\eta^{(k)}, y^{(k)}, q^{(k)}, d^{(k)}, \delta^{(k)})\}_{k=0}^{K}$ in Algorithm 11, and subsequently analyze the complexity of doing so.

Algorithm 11 Trace estimation procedure for the final solution

---

Input: Access to an s-sparse matrix $A \in \mathbb{R}^{n\times n}$ with $\|A\|_F \le 1$, state preparation pair description
  of solution $\{(\eta^{(k)}, y^{(k)}, q^{(k)}, d^{(k)}, \delta^{(k)})\}_{k=0}^{K}$, precision $\theta \in (0,1)$, $\zeta = \left(\frac{\epsilon}{n\|C\|_F}\right)^4$

Output: A θ-precise classical estimate of $\mathrm{tr}(A\tilde{\rho})$

Initialize: $a \leftarrow 0$, $k \leftarrow 0$, $y^{(K+1)} \leftarrow (0,0)^{\top}$, $\delta^{(K+1)} = 0$, $\eta^{(K+1)} = \frac{\sum_{k\in[K]}\eta^{(k)}(1+n\delta^{(k)})}{n\sum_{k\in[K]}\delta^{(k)}}$,

$Q^{(K+1)} \leftarrow ee^{\top}$

for $k = 0, \ldots, K+1$ **do**

  1. Implement an $(\alpha, a, \zeta/2(K+2))$-block-encoding of $Q^{(k)} \circ A$

  2. Use block-encoding of $Q^{(k)} \circ A$ to implement a trace estimator for

  $$a^{(k)} = \mathrm{tr}\left[\left(Q^{(k)}\circ A\right)\left(\frac{\exp\left(-\left[y_1^{(k)}Q^{(k)}\circ\widetilde{C} + y_2^{(k)}\,\mathrm{diag}\left(d^{(k)}\right)\right]\right)}{\mathrm{tr}\left(\exp\left(-\left[y_1^{(k)}Q^{(k)}\circ\widetilde{C} + y_2^{(k)}\,\mathrm{diag}\left(d^{(k)}\right)\right]\right)\right)}\right)\right]$$

  3. Use $\mathcal{O}\left(\frac{K}{\theta}\right)$ samples from the trace estimator to produce $\frac{\theta}{K+2}$-precise estimate $\tilde{a}^{(k)}$ of $a^{(k)}$

  4. Update solution:
  $$a \leftarrow a + \frac{1}{\eta^{(k)}(1 + n\delta^{(k)})}\tilde{a}^{(k)}$$

  5. $k \leftarrow k+1$

end

Theorem 7.25. *Let $A \in \mathbb{R}^{n \times n}$, and $\widetilde{C} \in \mathcal{S}^n$ be stored in QRAM, $\theta \in (0,1)$, and*

$$\{(\eta^{(k)}, y^{(k)}, q^{(k)}, d^{(k)}, \delta^{(k)})\}_{k=0}^{K}$$

be a state preparation pair description of the solution obtained from running Algorithm 10 to final precision $\zeta = \left(\frac{\epsilon}{n\|C\|_F}\right)^4$. Suppose A is an s-sparse matrix with $\|A\|_F \leq 1$, and assume classical access to A and $\widetilde{C} \in \mathcal{S}^n$. Then, Algorithm 11 outputs a θ-precise estimate of $\mathrm{tr}(A\tilde{\rho})$ using at most

$$\widetilde{\mathcal{O}}_{n,\|C\|_F,\frac{1}{\epsilon}}\left(\frac{\sqrt{n}}{\theta}\right)$$

queries to the QRAM and $\widetilde{\mathcal{O}}_{n,\|C\|_F,\frac{1}{\epsilon}}(ns)$ classical operations.

Proof. We begin by establishing the correctness of Algorithm 11. First, note that following the proof of Proposition 7.20, we can simplify the expression of the final solution to

$$\tilde{\rho} = \left[\sum_{k=0}^{K+1} \frac{1}{\eta^{(k)}(1 + n\delta^{(k)})} Q^{(k)} \circ \frac{\exp\left(-\left[y_1^{(k)} Q^{(k)} \circ \widetilde{C} + y_2^{(k)} \mathrm{diag}\left(d^{(k)}\right)\right]\right)}{\mathrm{tr}\left(\exp\left(-\left[y_1^{(k)} Q^{(k)} \circ \widetilde{C} + y_2^{(k)} \mathrm{diag}\left(d^{(k)}\right)\right]\right)\right)} \right].$$

by setting $y^{(K+1)} = (0,0)^\top$, $\delta^{(K+1)} = 0$, $\eta^{(K+1)} = \frac{1}{n}\sum_{k \in [K]} \frac{\eta^{(k)}(1+n\delta^{(k)})}{\delta^{(k)}}$, and $Q^{(K+1)} = ee^\top$. Then, by linearity of the trace and Lemma 7.1, one has:

$$
\begin{aligned}
\mathrm{tr}(A\tilde{\rho}) &= \mathrm{tr}\left(A \left[\sum_{k=0}^{K+1} \frac{1}{\eta^{(k)}(1 + n\delta^{(k)})} Q^{(k)} \circ \frac{\exp\left(-\left[y_1^{(k)} Q^{(k)} \circ \widetilde{C} + y_2^{(k)} \mathrm{diag}\left(d^{(k)}\right)\right]\right)}{\mathrm{tr}\left(\exp\left(-\left[y_1^{(k)} Q^{(k)} \circ \widetilde{C} + y_2^{(k)} \mathrm{diag}\left(d^{(k)}\right)\right]\right)\right)} \right] \right) \\
&= \sum_{k=0}^{K+1} \frac{1}{\eta^{(k)}(1 + n\delta^{(k)})} \mathrm{tr}\left(A \left[Q^{(k)} \circ \frac{\exp\left(-\left[y_1^{(k)} Q^{(k)} \circ \widetilde{C} + y_2^{(k)} \mathrm{diag}\left(d^{(k)}\right)\right]\right)}{\mathrm{tr}\left(\exp\left(-\left[y_1^{(k)} Q^{(k)} \circ \widetilde{C} + y_2^{(k)} \mathrm{diag}\left(d^{(k)}\right)\right]\right)\right)} \right] \right) \\
&= \sum_{k=0}^{K+1} \frac{1}{\eta^{(k)}(1 + n\delta^{(k)})} \mathrm{tr}\left(\left(Q^{(k)} \circ A\right) \frac{\exp\left(-\left[y_1^{(k)} Q^{(k)} \circ \widetilde{C} + y_2^{(k)} \mathrm{diag}\left(d^{(k)}\right)\right]\right)}{\mathrm{tr}\left(\exp\left(-\left[y_1^{(k)} Q^{(k)} \circ \widetilde{C} + y_2^{(k)} \mathrm{diag}\left(d^{(k)}\right)\right]\right)\right)} \right).
\end{aligned}
$$

In other words, the output of Algorithm 11 is indeed an estimate of $\mathrm{tr}(A\tilde{\rho})$.

Next, we analyze the complexity of the procedure. If A is classically known, one can store $Q^{(k)} \circ A$ in the QRAM using $\mathcal{O}(ns)$ classical operations, as A is s-sparse. With $Q \circ A$ stored in a QRAM data structure, one can apply Lemma 3.3 to implement an $(1, \log(n) + 2, \zeta/2(K+2))$-block-encoding of $Q \circ A$ in time $\widetilde{\mathcal{O}}_{\frac{nK}{\zeta}}(1)$ (as $\|Q \circ A\|_F \leq \|A\|_F \leq 1$ for any Q defined according to

(7.11)). As we saw in the proof of Proposition 7.20, with $\widetilde{C}$ stored in QRAM, one can implement the state

$$\rho^{(k)} = \frac{\exp\left(-\left[y_1^{(k)} Q^{(k)} \circ \widetilde{C} + y_2^{(k)} \operatorname{diag}\left(d^{(k)}\right)\right]\right)}{\operatorname{tr}\left(\exp\left(-\left[y_1^{(k)} Q^{(k)} \circ \widetilde{C} + y_2^{(k)} \operatorname{diag}\left(d^{(k)}\right)\right]\right)\right)}$$

using at most

$$\widetilde{\mathcal{O}}_n\left(\sqrt{n}\right),$$

accesses to the QRAM and $\mathcal{O}(n)$ classical operations.

Having prepared the state $\rho^{(k)}$ and a $(1, \log(n)+2, \zeta/2(K+2))$-block-encoding U_k of $Q^{(k)} \circ A$, Lemma 3.19 asserts that one can implement a trace estimator for

$$\operatorname{tr}\left[\left(Q^{(k)} \circ A\right) \rho^{(k)}\right]$$

with bias at most $\frac{\zeta}{K+2}$ using $\widetilde{\mathcal{O}}(1)$ applications of U_k and $U_k^\dagger$. Applying amplitude estimation using $\mathcal{O}\left(\frac{K}{\theta}\right) = \widetilde{\mathcal{O}}_{n, \|C\|_F, \frac{1}{\epsilon}}\left(\frac{1}{\theta}\right)$ samples from the estimator, we obtain a $\frac{\theta}{K+2}$-precise classical estimate $\tilde{a}^{(k)}$ of $a^{(k)}$, as $K = \mathcal{O}\left(\operatorname{polylog}\left(n, \|C\|_F, \frac{1}{\epsilon}\right)\right)$.

From here, we classically update a using $\mathcal{O}(1)$ arithmetic operations. Therefore, each iteration of Algorithm 11 requires at most

$$\widetilde{\mathcal{O}}_{n, \frac{K}{\zeta}}\left(\frac{\sqrt{n}}{\theta}\right)$$

accesses to the QRAM and $\mathcal{O}(ns)$ classical operations. Summing over $K + 2$ iterations implies a total of

$$\widetilde{\mathcal{O}}_{n, \frac{K}{\zeta}}\left(K\left(\frac{\sqrt{n}}{\theta}\right)\right) = \widetilde{\mathcal{O}}_{n, \|C\|_F, \frac{1}{\epsilon}}\left(\frac{\sqrt{n}}{\theta}\right)$$

accesses to the QRAM and

$$\mathcal{O}\left(Kns\right) = \widetilde{\mathcal{O}}_{n, \|C\|_F, \frac{1}{\epsilon}}\left(ns\right)$$

classical operations. The proof is complete. $\qquad\square$

Note that if $\|A\|_F > 1$, because of the subnormalization to block-encode A we need to increase precision of the estimation procedure: the cost increases by a factor proportional to $\|A\|_F$.

Corollary 7.5. *Let $A \in \mathbb{R}^{n \times n}$, $\theta \in (0, 1)$, and $\{(\eta^{(k)}, y^{(k)}, q^{(k)}, d^{(k)}, \delta^{(k)})\}_{k=0}^{K}$ be a state preparation pair description of the solution obtained from running Algorithm 10 to final precision $\zeta = \left(\frac{\epsilon}{n\|C\|_F}\right)^4$. Suppose A is an s-sparse matrix with $\|A\|_F \leq 1$, and assume sparse oracle access to A and $\widetilde{C} \in \mathcal{S}^n$. Then, Algorithm 11 outputs a θ-precise estimate of $\operatorname{tr}(A\tilde{\rho})$ using at most*

$$\widetilde{\mathcal{O}}_{n, \|C\|_F, \frac{1}{\epsilon}}\left(\frac{n^{2.5} s^2}{\theta}\right)$$

queries to O_A, O_C, and $\widetilde{\mathcal{O}}_{n, \|C\|_F, \frac{1}{\epsilon}}\left(\frac{n^{3.5} s^2}{\theta}\right)$ additional gates.

Proof. Provided classical access to A, we use Lemma 3.2 with $s_r = s_c$ to construct an $(s, \log(n) + 3, \theta/n)$-block-encoding of A with two uses of O_A (an oracle describing the elements of A in binary), and additionally using $\tilde{\mathcal{O}}_n(1)$ one and two qubit gates.

Likewise, with access to the oracle O_C describing the elements of $\tilde{C}$, one can construct an $(s, \log(n) + 3, \theta/n)$-block-encoding of $\tilde{C}$ with two uses of O_C, and additionally using $\tilde{\mathcal{O}}_n(1)$ one and two qubit gates. Note that without access to QRAM, we must compute the Hadamard products by taking the Hadamard products of block-encodings, which causes the subnormalization factor for the Hadamard product $Q^{(k)} \circ \tilde{C}$ to be ns, as $Q^{(k)}$ may be fully dense and C is s-sparse. It follows that preparing one copy of each Gibbs state requires

$$\tilde{\mathcal{O}}_n\left(\sqrt{n}(ns)\right) = \tilde{\mathcal{O}}_n\left(n^{1.5}s\right)$$

accesses to block-encodings of $Q^{(k)} \circ \tilde{C}$ and D, which each require an additional $\tilde{\mathcal{O}}_n(n)$ gates (to construct sparse-access oracles for $Q^{(k)}$ and D).

Similarly, the subnormalization factor for a block-encoding U_k of $Q^{(k)} \circ A$ will be ns. Having prepared the state $\rho^{(k)}$ and a block-encoding $Q^{(k)} \circ A$, Lemma 3.19 asserts that one can implement a trace estimator for

$$\mathrm{tr}\left[\left(Q^{(k)} \circ A\right)\rho^{(k)}\right]$$

with bias at most $\frac{\zeta}{K+2}$ using $\tilde{\mathcal{O}}(ns)$ applications of U_k and $U_k^\dagger$. Applying amplitude estimation using $\mathcal{O}\left(\frac{K}{\theta}\right) = \tilde{\mathcal{O}}_{n,\|C\|_F,\frac{1}{\epsilon}}\left(\frac{1}{\theta}\right)$ samples from the estimator to obtain a $\frac{\theta}{K+2}$-precise classical estimate $\tilde{a}^{(k)}$ of $a^{(k)}$, as $K = \mathcal{O}\left(\mathrm{polylog}\left(n, \|C\|_F, \frac{1}{\epsilon}\right)\right)$.

Just as in the QRAM setting, classically updating a requires $\mathcal{O}(1)$ arithmetic operations. Therefore, without access to QRAM, each iteration of Algorithm 11 requires at most

$$\tilde{\mathcal{O}}_{n,\frac{K}{\zeta}}\left(\frac{n^{2.5}s^2}{\theta}\right) = \tilde{\mathcal{O}}_{n,\|C\|_F,\frac{1}{\epsilon}}\left(\frac{n^{2.5}s^2}{\theta}\right)$$

applications of block-encodings for $Q^{(k)} \circ \tilde{C}$, $D^{(k)}$ and $Q^{(k)} \circ A$ and $\tilde{\mathcal{O}}_{n,\|C\|_F,\frac{1}{\epsilon}}\left(\frac{n^{3.5}s^2}{\theta}\right)$ additional gates. This corresponds to $\tilde{\mathcal{O}}_{n,\|C\|_F,\frac{1}{\epsilon}}\left(\frac{n^{2.5}s^2}{\theta}\right)$ queries to O_A and O_C in each iteration, and $\tilde{\mathcal{O}}_{n,\|C\|_F,\frac{1}{\epsilon}}\left(\frac{n^{3.5}s^2}{\theta}\right)$ additional gates. Summing over the $K + 2 = \tilde{\mathcal{O}}_{n,\|C\|_F,\frac{1}{\epsilon}}(1)$ iterations yields the stated complexity. $\qquad\square$

www.ingramcontent.com/pod-product-compliance
Lightning Source LLC
LaVergne TN
LVHW020745200726

843506LV00009B/888